Debojyoti Halder
Asim Sarkar

Astronomical Studies on SETI

Debojyoti Halder
Asim Sarkar

Astronomical Studies on SETI

LAP LAMBERT Academic Publishing

Imprint
Any brand names and product names mentioned in this book are subject to trademark, brand or patent protection and are trademarks or registered trademarks of their respective holders. The use of brand names, product names, common names, trade names, product descriptions etc. even without a particular marking in this work is in no way to be construed to mean that such names may be regarded as unrestricted in respect of trademark and brand protection legislation and could thus be used by anyone.

Cover image: www.ingimage.com

Publisher:
LAP LAMBERT Academic Publishing
is a trademark of
Dodo Books Indian Ocean Ltd. and OmniScriptum S.R.L publishing group

120 High Road, East Finchley, London, N2 9ED, United Kingdom
Str. Armeneasca 28/1, office 1, Chisinau MD-2012, Republic of Moldova, Europe
Managing Directors: Ieva Konstantinova, Victoria Ursu
info@omniscriptum.com

Printed at: see last page
ISBN: 978-620-8-42637-8

Astronomical Studies on SETI

Debojyoti Halder and **Asim Sarkar**

Department of Physics,
R. B. C. Evening College,
Naihati, West Bengal, India
debojyoti_halder@rbcec.in

Debagram S. A. Vidyapith (HS)
Debagram, Kaliganj,
Nadia, West Bengal, India
asim693@gmail.com

Dedicated to our Parents

PREFACE

This book is mainly based on the doctoral thesis entitled "*Interstellar Studies from Reception of Radio Signals Using Hydrogen Line Spectrometer and to Investigate the Transient Emissions from the Sun*". We have modified some chapters and added some new material for the sole purpose of this book. The references used for this book are also given at the end of each chapter. A glossary of scientific terms used is added as well for improvement of easy understanding of related subjects.

Search for extraterrestrial intelligence (SETI) is indeed a great challenge to astronomers since its method of communication, direction of propagation as well as spectral pattern are all unknown earlier. Due to all of these various restrictions, the problems have become complex and thus the science has become a 'hard science'. As galactic societies would be transitory, an obvious solution is an interstellar communication network, or type of library mostly consisting of automated systems to store the cumulative knowledge of vanished civilizations and thereby to communicate the knowledge through the galaxy. This is called 'interstellar internet', where the different automated systems act as network servers. Intercepting such signals is highly difficult and extremely complex. The article reviews the modern techniques and instruments for celestial search. Theories and hypothesis developed for investigating life in the universe including some exciting findings have been critically examined.

In recent years, Scientists from different parts reported the interaction between the Sun and the Earth. It is proved by various means that there is a great influence of the Sun on the Earth's atmosphere. The success of radio signal transmission and reception largely depends on the position where to look in the sky, besides frequency, bandwidth, polarization and also on the type of modulation study. The Sun's variable magnetic fields constitute a rich source for processes that influence the Earth's upper atmosphere and geomagnetism. The Sun follows a periodic cycle of activity. This cycle, called the solar cycle, has its evident manifestation by the periodic recurrence of sunspots, or darker, relatively cool and strong magnetic regions at the Sun surface. The solar cycle usually lasts around eleven years on the average and there is little doubt that it is magnetic in nature and produced by dynamo processes within the Sun.

The explosions (flares and coronal mass ejections) on the Sun can cause highly energetic particles and millions of tons of very hot, electrified material to be expelled, sometimes on a collision course with Earth. This material also carries magnetic field from the Sun which is very important when we are considering how the material will affect the Earth. If this material manages to break its way into the Earth's normally protective atmosphere (the magnetosphere) we can see many effects on Earth. The bi-polar magnetic field of the Earth points north but the field contained within the material expelled from the Sun can point in any direction. When the field is orientated in the opposite direction to that of the Earth, that is when it points south, the two magnetic systems interact and the solar material can enter the

Earth's magnetosphere. Established methods for detecting exoplanets are briefly discussed. Radio method implemented for early searches of exoplanets is then critically focused showing the confirmed exoplanets up to recent time. Our approach is to receive and transmit radio signals by considering the two "water holes" and the associated frequencies. Under certain approximation the probable numbers of stars that can be covered by the proposed radio signal technique have been determined, pointing out the limitations. We have taken into account the available data of known stars/constellations up to the recent years to find the required time of reception of the echo radio signals, if transmitted using the wavelengths corresponding to the "water holes" from the earth's station. It further investigates the probable responding number of stars due to different transmitted power. Utilizing the Drake equation two extreme boundaries have been first determined from the conservative and optimistic point of view and thereby extracting the realistic parameters the probable number of broadcasting civilizations has been estimated. The limitation of the equation is outlined and its suitable modifications have proposed.

A conceptual diagram of the terrestrial global circuit is given from the existing knowledge of earth's electrical system. When the reported results are critically reviewed Mars is also expected to have substantial atmospheric electrification. The spontaneous leakage of artifacts into the interstellar medium and the associated orbital characteristics of different planets have been considered for estimating the frequency of the artifacts likely to fall down to the Earth and the other planets. Information about the astro-dynamical aspect of paleohistology, particularly regarding the formation and extent of the artifacts in the solar planetary systems has been emphasized. A significant geomagnetic activity during major meteor showers indicates a correlation between the Orionid meteor and the magnetic phenomena. Our analysis suggests that highly ionized debris of meteors and absence of photo-detachment process are responsible for attracting meteor effects on radio signals more at night than that at daytime. The meteorology of the high-altitude observing station Darjeeling including the GPS data taken at the observatory has been presented. The Jove receiver used for the reception of radio signals and the technique employed for the detection of bursts are reported.

January 10, 2025 — *Debojyoti Halder*

West Bengal, India — *Asim Sarkar*

CONTENTS

About the Authors

Dr. Debojyoti Halder

Dr. Debojyoti Halder completed his Ph. D. in 2016 from the University of Kalyani in the applied field of Physics. He has designed and constructed Radio Astronomical Instruments which includes wind-proof Log Periodic Dipole Array for investigating the dynamic behavior of the Sun. He has already published more than 25 research papers in reputed International and National Journals. His field of interest includes Radio Astronomy, and Communication Applications. Presently, he is working as an Assistant Professor and Head of Department of Physics for last 9 years in Rishi Bankim Chandra Evening College, India. He wishes to continue research in the aforesaid fields in future.

Dr. Asim Sarkar

Dr. Asim Sarkar obtained his Ph. D. degree in 2015 from the University of Kalyani in the Applied field of Physics. He has worked in the field of Radio Astronomy and his research work was mainly to investigate the radio signals coming from Interstellar space through the Water Hole at 1420 MHz frequency. His valuable research papers have been published in many reputed National and International journals. Always interested in learning about the mysteries of space, Dr. Sarkar is now enjoying his career as an Assistant Teacher in a Govt Sponsored School in West Bengal, India.

....

Chapter 1

BACKGROUND OF THE STUDY

1.1 Introduction

In recent years scientists have realized that intelligent life may well be found throughout the universe and we are, perhaps, not the only civilization in our galaxy. The work suggests an alternative search strategy through sky noise windows emphasizing both the “water holes” and the associated wavelengths. It also suggests that instead of pointing the antenna at known stars, an all-sky search would be more appropriate for interstellar communication. Both impulsive and gradual energy release take place from the different activity centers of the sun. The energy release processes have their signatures in the form of bursts in the radio frequencies. Depending upon the energy release and the transfer processes the solar radio bursts may cover a large bandwidth and at the same time may have fine frequency structure [1]. This article intends to inform about the history of radio astronomy, its evolution, connection with solar studies, and the contribution of the spacecraft will have on the study of this field [2, 3].

1.2 Some Earlier Phenomena at the Beginning of Radio Astronomy

It is almost impossible to depict the most important facts in the history of radio astronomy without presenting a sneak-peek where everything started, the development and understanding of the electromagnetic spectrum. In 1960 James Clerk Maxwell developed his famous theory related to electro-magnetic radiation and shown that it is associated with electricity and magnetism, and could exist at any wavelength beyond the visible ray region. Scientists then used radio signals in spite of optical signals for detecting solar signals because optical signals may be plagued by clouds and dusts present in earth's atmosphere. Radio signals have become one of the most important tools of astronomical observations [4, 5]. The earliest experiments in radio astronomy were not successful, mostly due to technical limitations of the instruments as well as lack of knowledge of the nature of radio emissions from celestial objects. Several attempts were made to detect radio emission from the Sun by experimenters of Oliver Lodge, Nikola Tesla and Thomas Edison but unfortunately, they were not successful [6, 7].

1.3 First Victory in Radio Astronomy

The earliest achievement was accidental! In 1932, physicist and radio engineer Karl Jansky was studying the origins of radio frequency interference for Bell Laboratories. The Bell Telephone Company was having problem with the working of their transatlantic service, because of static of some sort in the late 1930s. The company asked Karl Jansky to find such interference source [8]. He discovered a one type of interference signal (celestial in nature) with his Bell Laboratories ship, varying with the rise and set of constellations, coming from an unknown source in the center of our galaxy (Figure 1.1). He found "...a steady hiss type static of unknown origin", at a frequency of around 20 MHz which eventually he concluded had an extraterrestrial origin [9]. This was the first time that radio waves were detected from outer space. Eventually, through experimentation, he concluded that the signal received by him was not noise but was coming from the center of the Milky Way [10-12].

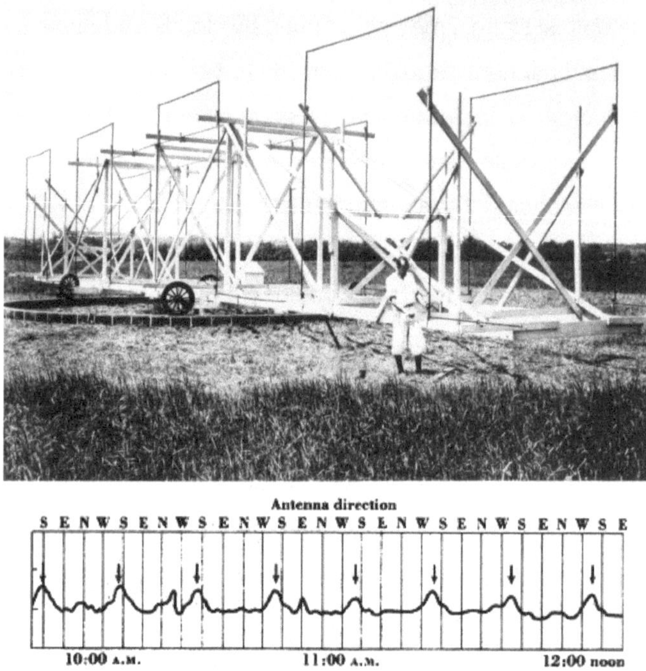

Figure 1.1 Karl Jansky detected radio waves coming from an unknown source in the center of our galaxy [9]

The first radio telescope was constructed by Grote Reber in 1944. He built a parabolic sheet metal mirror 9 meters in diameter, focusing to a radio receiver 8 meters above the mirror. This instrument is allowed to be pointed in various directions in the sky [13]. He is the first engineer who constructed the first celestial radio map of the sky. Though Jansky and Reber published their findings in many appropriate journals, their findings were ignored initially many times. The interest is begun after World War II; when modern technology had been developed rapidly. In particular accelerated development in radar and radio communications meant that in 1945 there was an abundance of surplus equipment and qualified staff in radio engineering world-wide.

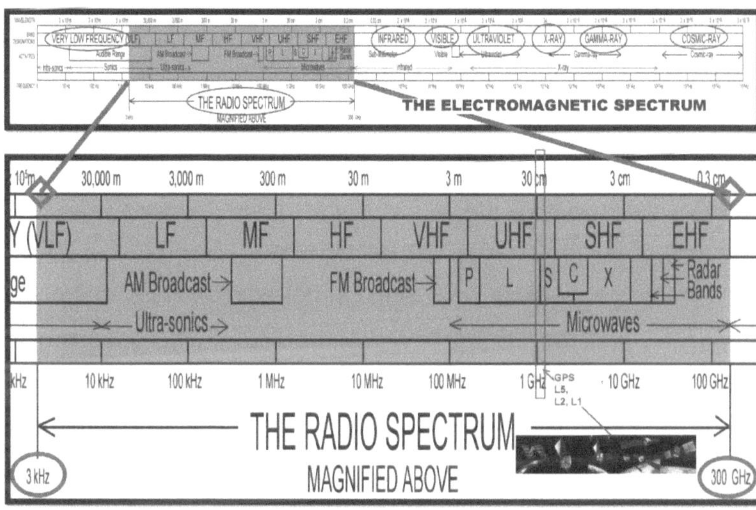

Figure 1.2 From light waves confined to a "window" about one octave wide, Jansky opened a broader radio "window" dozens of octaves wide.

1.4 World War II and the Explosion of Radio Astronomy

Radar had begun to play a vital role in the Second World War and afterwards. Range and strength of transmitters, and the reliability and sensitivity of receivers also provided an advantage. A stronger signal could travel farther, and a more sensitive receiver could pick up signals farther away. The technique of bouncing signals off the stratosphere solved the problem requiring a direct line of sight between transmitter and receiver. Radio-physics group was formed and given responsibility for research into radar. Radio physics played a vital part in creating mobile radar stations called light weight air-warning tents, small and reliable enough to make transportation by the allied forces during the war worthwhile [9, 14-16].

1.5 Radio Signals using Hydrogen Line Spectrometer

Astronomical radio sources are objects in outer space that emit strong radio waves. Radio emission comes from a wide variety of sources. Such objects represent some of the most extreme and energetic physical processes in the universe [17-18]. In the 1970s, some stars in our galaxy were found to be radio emitters, one of the strongest being the unique binary MWC 349. Some astronomical radio sources are Sun, galactic center, Supernova remnants, Pulsars, Star forming regions, Radio galaxies, Cosmic microwave background, Primordial black holes, Extragalactic pulses etc. [19-21].

The Dutch astronomer Jan Oort had spent many years studying the rotation and structure of the galaxy using optical means but he was frustrated by the extensive clouds of dust lying in the galactic plane, which block visible light [22-24]. One can see only a few thousand light years towards the galactic center because the light of distant stars is absorbed. After learnt Reber's discoveries in radio astronomy, Jan Oort realized that a radio spectral line would be an important tool for discovering the structure of our galaxy. Actually, radio waves can penetrate the dust to show the galactic center and indeed the opposite side of the galaxy. The importance of a spectral line is that the frequency of the line will be shifted due to Doppler

Effect which means that the velocity of the gas can be measured. One can then study the differential rotation of the galaxy and estimate distances to gas clouds and thus map the distribution of matter in the galaxy [25-26]. Jan Oort assigned his student Hendrik Van de Hulst, with the job of figuring out what radio spectral lines might exist and what their frequencies would be [27]. In 1944, De Hulst predicted the existence of a hydrogen emission line at 21.2 cm based on the physics of the hydrogen atom. The electron in a neutral atomic hydrogen atom may undergo a “spin- flip transition” where the spin property of its electron changes to be opposite to the proton in the atom, bringing the atom to a lower energy state and emitting a photon. The transition is rare for a single hydrogen atom, on average occurring once every 11 million years, but for the vast quantities of hydrogen that exist in a Galaxies and interstellar clouds, there are enough transitions to emit a steady signal. The original paper by Van de Hulst predicting the existence of the 21 cm line expressed doubt that the line would be detectable. However, a paper by Shklovski in 1948 was more optimistic [28, 29]. Since hydrogen is the most abundant element in the universe, he started his studies with hydrogen. He found that a "hyperfine" transition in the ground state of neutral hydrogen would produce radiation in the radio range, at a frequency of 1420 MHz i.e. at a wavelength of about 21 cm. In the ground state of hydrogen, the electron can have its magnetic moment either parallel to that of the proton or anti-parallel. The parallel state has a little more energy and so a transition to the anti-parallel state results in emission of 21 cm radiation.

The first successful detection was by Ewen and Purcell at Harvard. In 1950 Ewen and Purcell attempted to build a receiver for detecting the 21 cm line of neutral hydrogen [30-32]. Ewen and Purcell assumed that the Dutch were probably not working on detecting the line and that the Russians might soon be. They could probably make the first sensitive detection attempt. It seemed likely that the experiment would yield a negative result but the experience would be worth the trouble and a well-defined upper limit to the 21 cm line emission strength would be important. Ewen proceeded to design the horn antenna, the mixer and the receiver consulting with experts in these fields, including Sam Silver on antenna design and Bob Pound on mixers. The receiver used a frequency switching technique with a view to cancel out background noise, a novel technique for astronomy at the time. The Dutch group implemented Ewen's frequency switching technique and succeeded in detecting the line. A further confirmation of the detection was made by Australian radio astronomers [33]. In the following years, observations of the neutral hydrogen line were made with increasing accuracy by many astronomers, primarily Dutch, American, and Australian. The technique proved to be successful in mapping the large-scale distribution of matter in the galaxy [34-35].

In recent years, radio communication has been attempted by considering the sky noise windows of the atmosphere [36-38]. If the hydrogen line and the water line are taken into account one can find that the hydrogen (H) line is at 21 cm whereas the hydroxyl radical (OH) line is at 18 cm. When hydrogen is combined with hydroxyl radical, it produces water (H_2O) whose wavelength lies between 18 cm and 21 cm. The window between 18 cm and 21 cm is treated as the first 'water hole' which when converted to frequency lies between 1420 MHz to 1666 MHz.)

1.6 Practical Implications of the 21cm Hydrogen Line

Until the discovery of the hydrogen line there was almost no interest in radio emissions of wavelength below 1 meter. Shorter wavelength observations require more precise surfaces on the receiving dish, increasing the cost of engineering and manufacture. Furthermore, if a mesh is used in place of a solid surface the mesh must be finer to reflect radio waves of shorter wavelengths. The Jodrell Bank Mark I telescope, was initially designed outside tolerances that would allow 21cm observations. It was "far too rough for observations at 21 cm radio waves reflected from the surface would arrive at the focus in a scrambled, incoherent fashion." [9]. Mark I was redesigned during early construction [39-41]. The 100ft (30.5m) central section was replaced with a mesh of finer weave, and made more rigid to conform to the desired shape within tolerances allowing 21cm waves to arrive at the receiver coherently. Design changes on a large structure tend to cascade. In this case the back girder and support towers had to be strengthened to support the extra weight and wind loading. The existence of the 21cm line had been confirmed in 1951, and excavation for the Mark I began in 1952. It is therefore unsurprising that this redesign was necessary. Had the H- line been discovered later, changes required may not have been feasible, Making Mark I less useful than it turned out to be [9]. The cost of Mark I escalated causing its designers and builders financial and legal problems. Their Australian counterparts were determined not to repeat the mistake with their giant radio telescope [42-44].

The Parkes telescope was designed from the start with a surface accurate enough for observations down to 10 cm, ensuring that it would perform well at 21 cm. The aim was to build a bigger better dish than the Mark I. When the insufficiency of fund for a larger dish was realised, parkes' designers focused on making a better performing instrument [45]. A parabolic steerable dish of maximum gain was exactly what was needed for galactic H- line work [46-47].

1.7 Early Works on Hydrogen Line

The initial 1951 observation of the hydrogen line had taken Ewen and Purcell almost two years of work to achieve. Two groups set about to confirm the discovery, each taking about 2 months to do so. The Sydney group consisted of Chris Christiansen and Jim V. Hindman at the Potts Hill field station, who set aside solar studies to assemble the necessary equipment and confirm the Harvard result as soon as they were made aware of it by a cable sent by Frank J. Kerr. (Frank Kerr, based in Sydney at Radio physics, was in Harvard for a year undertaking an astronomy course.) The Leiden group consisted of Jan Oort, who had been involved in refining predictions of the hydrogen line and was respected for a variety of other works, and Alexander Muller. At Purcell's suggestion, the papers confirming the discovery of the hydrogen line appeared along with the original paper by Ewen and Purcell on the same issue of 15 *Nature* [9, 30, 32, 48, 49]. With the existence of the hydrogen line confirmed, it was a natural next step for Christiansen and Hindman to carry out observations of the sky along the plane of the Milky Way. Within 3 months of their initial confirmation of the existence of the hydrogen line, they had found evidence of a spiral structure in our Galaxy [50, 51].

Grote Reber was motivated by Jansky's discovery. He was interested to find out the process that make the development of radio waves in space and verified whether the waves were coming from the different celestial objects. Bell Labs were not so successful at that time of great depression. Reber had the determination to achieve his goals, even if he did it all from his back yard. On his own investigation Reber constructed a telescope in 1937 which had a parabolic dish reflector and 3 receiving frequencies: 3300MHz, 900MHz and 160MHz [52]. In 1938, the last receiver gave him the most wanted galactic radio waves. Reber's data was shown as contour maps which represented the Milky Way with bright areas [53]. Reber was one of the pioneers of modern radio astronomy. Depending on his work, after World War II, many scientists were able to build huge and better antennas or studying the universe [54].

In these days, we have big radio telescopes like the Arecibo Radio telescope in Puerto Rico (Figure 1.3). It has 305m diameter and 167 feet depth and can cover an area of about twenty acres [55]. A technique like radio interferometry was available in 1946. Using multiple antennas Very Long Baseline Interferometry (VLBI) and the Space Very Long Baseline Interferometry (SVLBI) are some sophisticated techniques to record radio data. Projects like the JPL SVLBI use space antenna technique to provide "4 to 10 times the resolution of VLBI" [56]. Improved radio astronomy field can detect the radio emissions from planet, energetic objects like quasars, pulsars, and radio galaxies made possible [57, 58]. Giant Metrewave Radio Telescope (GMRT), located near Pune in India, is an array of radio telescopes at meter wavelengths (Figure 1.3). It is operated by the National Centre for Radio Astrophysics (NCRA), a part of the Tata Institute of Fundamental Research, Mumbai. At the time it was built, it was world's largest interferometric array [59-60]. Chronology in Radio Astronomy at a Glance in presented in a Table 1.1.

Figure 1.3 Arecibo Radio telescope in Puerto Rico [55]

Figure 1.4 An antenna of the Giant Metrewave Radio Telescope (GMRT)

Table 1.1 Chronology in Radio Astronomy at a Glance [61]

Year	Description
1860-1870	James Clerk Maxwell developed his famous theory related to electro-magnetic radiation and shown that it is associated with electricity and magnetism and predicts that light will prove to be only a small part of a spectrum of electromagnetic radiation.
1888	Hertz proves that electromagnetic radiation exists and generates radio waves in his laboratory

1890s	Scientist attempt unsuccessfully to detect radio waves from the sun
1901	Marconi developed first practical long-distance radio communication system
1928	Jansky begins working for Bell Telephone Laboratories and is assigned to find out what types of interference might block short radio waves
1931	Jansky recorded radio interference from different parts of the sky
1932	He discovered three type of radio sources of statics, including a hiss of unknown origin
1933	Jansky conclud that this unknown radio sources coming from the centre of the Milky Way Galaxy.
1937	Grote Reber Built first radio telescope
1938	Reber confirmed Jansky's observation of radio emission
1940	An article about radio astronomy is published in an astronomy journal for the first time
1941-1944	Reber made maps of the sky showing sources of radio wave emission

1945	Van de Hulst predicts that interstellar hydrogen will give off 21 cm radio waves
1951	Ewen and Purcell detect 21 cm hydrogen radiation from the Milky Way.

1.8 Radio Signals from Various Natural Sources

1.8.1 Radio Signals coming from the Sun

The Sun was one of the first objects studied by early radio astronomers. It is not as powerful an emitter of radio waves as many other objects, but its close proximity to us makes it appear radio-bright to us here on the third planet. In the year 2000, the Sun is expected to peak in sunspot number and the related solar activity level. That means there should be lots of solar flares on the Sun's surface and the Earth should receive a number of geomagnetic storms as a result. In other words, this should be an exciting time to begin monitoring the Sun with radio receivers and magnetometers [62-66].

1.8.2 The Galactic Center

The galactic center of the Milky Way was the first radio source to be detected. It contains a number of radio sources, including Sagittarius A* and the supermassive black hole at its center.

1.8.3 Radio Signals coming from the Jupiter

Mysterious signals from space were discovered in 1955 by radio astronomers [36, 67] and thought initially that the signals were local interference; some even considered it a noisy ignition system of a pickup truck. From a systematic study it was noted that the planet Jupiter was in the beam of the cross antenna at all the time when signals were heard and recorded. Unlike many radio astronomy dish antennas, the huge Mills Cross comprised more than 100 dipoles strung between wooden poles. The dipoles were phased for producing a narrow, steerable, pencil-thin beam with 2.5° in width, considering the operating frequency as 22.2 MHz. Researchers have aimed shortwave antennas at Jupiter with a view to understand the source of these powerful signals. Giant Jupiter is a large ball of gas with a small core of solid hydrogen and a strong magnetic field. The four large moons of Jupiter *viz.* Io, Europa, Ganymede and Callisto, were first viewed by Galileo utilizing his primitive telescope. Jupiter emits cyclotron radiation from electrons trapped; more specifically, in its strong magnetic field [68-70]. This range of emission varies from 1 MHz to ~39.5 MHz with a peak at about 10 MHz. However, the frequencies below ~15 MHz are unsuitable to penetrate the ionosphere of the Earth and again the intensity of the emissions drops off rapidly if we go at higher values. Consequently frequency ~20 MHz is taken as an optimal frequency for receiving Jovian emissions on the Earth's surface. After the reception of radio bursts from Jupiter, scientists tried to clarify the cause of this radio mission. They started observations, recording the time of the received signal and analyzed the intensity of the decametric radio bursts, particularly the part of Jupiter facing them at a certain time which can be achieved from its rotation rate. It was understood that audible sound from Jupiter largely depends on what part of Jupiter is facing us at that time. It was also realized that the radio emission depends on the Jovian longitude.

These longitudes are something like "landmarks" which mean that Jupiter is not only sending out radio waves in every direction but also is beaming the radio waves into space. Astronomers from careful observations were able to find changes in the location of the radio sources on Jupiter and the emissions behavior [71, 72]. Techniques employed for the detection of Jovian radio signal bursts are discussed and some preliminary results are reported.

1.8.4 Radio Signals coming from the Venus

Radio signals of an impulsive or burst-like nature which appear to come from the planet Venus have been detected at the Ohio State University at a wave-length of 11m. The observations were made with an interferometer antenna consisting of two colinear arrays of six horizontal half-wave-length elements one-quarter wave-length above ground. Each array is backed up by a reflector element, which, with the ground, forms a corner reflector. The arrays are separated by 426 ft. along an east–west axis. The receiver operates a pen recorder with a time constant of about one-quarter second [73]. In two earlier communications, the reception of impulsive radio signals from the planet Venus at a wave-length of 11 meters was reported. These signals have burst-like characteristics somewhat resembling atmospherics from terrestrial thunderstorms and may presumably originate in similar storms on Venus. The bursts are of very short duration, being a small fraction of a second.[74] The June 6, 2012 transit of Venus across the disk of the Sun reveals that there are a number of short periods fading in the atmospherics level during the period of transits of Venus [75]. The radio signal received at 92.7 MHz also exhibits a rapid variation of the voltage level with a sudden enhancement before the greatest position of the Venus. Our analysis suggests that the earth's ionosphere is affected during contact I to contact IV of the transits causing an effect on the ionospheric propagation [76-77].

1.8.5 Radio Signals coming from Natural Hazards (Earthquake)

VHF receiver at 92.7 MHz operating at Kalyani, West Bengal, recorded characteristic radio signal during the Sikkim Earthquake of 18 September, 2011. It reported the significant variations of the received signal and the results had been interpreted on the basis of electromagnetic signal perturbations affecting the lower ionospheric layers causing propagation between the arth-ionosphere waveguide. In association with the Indonesia earthquake followed by major cities of India, the characteristic variation in the atmospheric records at Kalyani using VLF atmospherics at 27 kHz [78-79].

1.8.6 Radio Signals coming from Meteor Showers

It is well known that a meteor can be a source of ionization in the upper atmosphere of the earth [80, 81]. The ionization in meteor burst is often harnessed for developing a communication link in the transmission, unaffected by sudden ionospheric disturbances [82, 83]. The same ionization may be generally responsible at different frequency bands during propagation of signals through it. The radio data obtained from distant objects like stars, galaxies, pulsars, meteors, artifacts etc. are useful for studying the Universe [84, 85]. An observed association of sawtooth fadeout (SF) effects in the field strength of VHF radio signals and also at 40 kHz VLF atmospherics with Orionid meteor showers had been reported, due to a significant geomagnetic activity occurred, during major meteor showers. It indicated a correlation between the Orionid meteor and the magnetic phenomena. It suggested that highly ionized debris of meteors and absence of photo-detachment process were responsible for attracting meteor effects on radio signals more at night than that at daytime [86].

1.9 Research on Solar Radio Wave

Scientists have proposed and determined that the emitted radio waves are the result of different solar activities. Though the mechanism of various solar activities still remains unknown, the results of them are quite significant [87, 88]. Some mysterious solar flares cause as a result of a magnetic sheer, for the kinking in the magnetic fields of the sun that in turn crosses each other. Radiations through the whole electromagnetic spectrum, including radio waves are emitted, when this happens. We are able to detect the waves lying in the electromagnetic spectrum's radio area. They would appear as a peak with a brief duration and significant amplitude in our data collection [89].

Transient astronomical electromagnetic radiation is the signature of some of the most fascinating physical phenomena in the universe. Anywhere the physical conditions are changed with time, there is the potential for transient radiation [90-92]. The detection of transient astronomical radiation presents challenges not encountered in observations of persistent sources. Sources which produce radiation sporadically cannot easily be studied with typical observatories and their schedules of observing time allocation; other techniques are required. Detecting transient astronomical signals at radio wavelengths in particular requires overcoming an even greater difficulty: the pervasive presence of radio frequency interference (RFI) [93-94]. The ever-growing use of wireless services means the radio spectrum is crowded with a wide variety of signals, a large fraction of which are transient. Since it is the very nature of transient signals of terrestrial or astronomical origin to disappear unpredictably, they cannot reliably be distinguished from each other by repeated observing. Surveyors for transient astronomical radio signals must devise other methods to separate the desired astronomical signals from the seemingly ubiquitous RFI [95-96].

1.10 Sources of Transient Astronomical Radio Emission

Many sources have been observed to produce transient astronomical radio emission. Perhaps the most familiar is the Sun, from which a large variety of transient radio signals emanate. From microwave spike bursts lasting 10 *ms* to Type III storms lasting weeks, the characteristics of solar radio bursts span wide ranges of brightness temperature, duration, frequency, and polarization. Associated with many different aspects of solar activity, the mechanisms producing the bursts run the gamut, from thermal bremsstrahlung to plasma radiation [96-97]. Other stars have been seen to produce radio bursts as well. Events observed on flare stars are similar in character to those seen on the Sun, but imply radio luminosities 10^4 times that of solar acres [99]. RS CVn systems lose binaries with orbital periods of 1-30 days, have been observed to produce transient radio emission on time scales of minutes to days.

The signals are thought to be due to two separate phenomena: the acceleration of electrons in the magnetic fields between the stars, and coherent emission from the individual stars [100-101].

The corona is only about one-millionth of the brightness of the Sun. As a matter of fact, its observation, outside eclipses, is quite difficult. Total solar eclipses are rare and also visible from very restricted areas of the Earth. Observing white-light corona using a coronagraph is a difficult task as it needs both a perfect instrument and an ideal site. However, monochromatic emission from the corona with less perfect observing conditions is possible with a suitable monochromatic filter, which is less costly than the interference filter developed around 1950. At the beginning of 1960's a small sized coronagraph suitable for amateurs was developed. Twenty-five years later the study of the corona has taken on new life with the development radio methods [102, 103].

1.11 The Solar Variability

Solar variability originates from the interior of the Sun. One very important aspect of solar variability is to link solar magnetic fields generated below the convective zone in the interior. The evolution of magnetic fields is responsible for variety of manifestations like the well-known coronal mass ejections. Solar activity changes on time-scales from hours to years. Broadly one can classify the two distinct types of emissions from the Sun as (i) continuous emission and (ii) solar cycle dependent emission. Electromagnetic radiation, solar wind and solar magnetic field are the continuous emissions, whereas coronal mass ejection, energetic particle emission are related to solar cycle dependent emission. The continuous emissions are caused of climate, structure of the atmosphere, magnetosphere. The solar cycle dependent emissions are the cause of geomagnetic activity and it interacts with upper polar atmosphere. One very important aspect of solar variability is linked to solar magnetic fields generated below the convective zone in the interior. The evolution of magnetic fields is responsible for variety of manifestations such as the well-known coronal mass ejections and solar wind structures. Though the influences from the variation of the electromagnetic radiation have considered widely and largely available in literature [104-106], a changing solar wind and energetic particles have got less priority, perhaps due to its very complex scenario, which can also influence Earth either directly or indirectly through modulating the cosmic ray flux. Variation of solar activity directly interacts with earth's atmosphere and also effect the climate of the earth. A block diagram (Figure 1.5) is shown the interaction between solar activities with earth's climate.

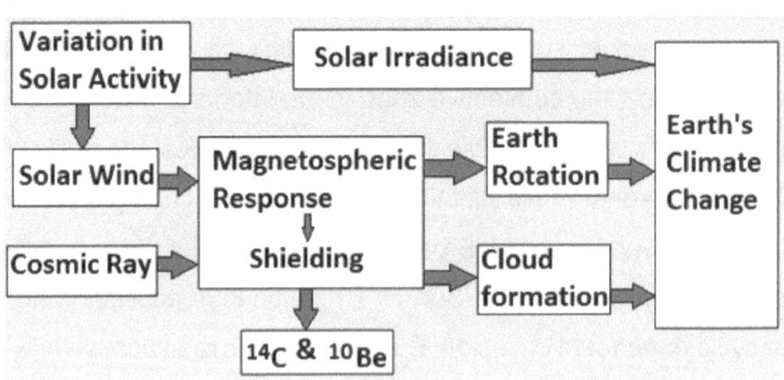

Figure 1.5 Change of Earth's climate with solar activity variation.

Variations in Solar activity led to changes in the Solar Wind and in solar irradiance, both of which may affect Earth's climate. The variations in irradiance are known to be small or even minute. The variations in Solar Wind are large and strong, via the interaction with the Earth's magnetosphere, it affects Earth's rate of rotation, and by that forcing several different terrestrial variables like the Gulf Stream beat in the North Atlantic. Simultaneously, the shielding capacity affects the concentration of cosmogenic nuclides [107], like the aa-index [108]. At any rate, there are two different ways for how changes in solar activity may affect Earth's climate; via irradiance or via the Solar Wind. In the present case, it is the left line that is considered.

1.12 Solar Coronal Mass Ejections

Coronal Mass Ejections (CMEs) originate deep in the corona and, due to physical processes that are currently not well understood, are often accelerated to speeds well above the average solar wind speed [109-115]. Solar wind compositional measurements provide important information about the source of such ejections, the heating profiles, and their expansion in the inner heliosphere. In this article we present data from the Solar Wind Ion Composition Spectrometer (SWICS), part of the Advanced Composition Explorer (ACE) spacecraft, describing the kinetic and compositional properties of the solar wind associated with CMEs that occurred in late October and early November 2003 [116-118].

First, the identification of ICME material has been the focus of many publications, e.g., by Richardson et al. [117, 119]. Common plasma signatures of ICMEs in the solar wind include counter-streaming electrons, anomalously low proton and electron temperatures, strong magnetic fields, low plasma beta, and smooth field rotations [120-121]. These studies have stressed the diverse nature of ICME-associated plasmas and the resulting lack of a coherent set of signatures that can be identified as “necessary and sufficient” for the presence of ICME plasma [122-123].

1.13 Variation of Earth’s Geomagnetic Field

Most of the geomagnetic field observed on the ground was of internal origin (About 99%) and a very small portion was of external origin. These external origins are known as transient variations. There is much kind of transient variations e. g. solar and lunar daily variations, magnetic storm, solar flare effects and solar coronal mass ejections (energetic charge particles). The primary cause of these variations is external to the earth. When we observed the geomagnetic variations, the variation is contained a part of internal origin, which is the secondary effect of the primary causes of same order of magnitude [124].

A geomagnetic storm is a temporary disturbance of the Earth's magnetosphere caused by a solar wind shock wave and/or cloud of magnetic field which interacts with the Earth's magnetic field. The increase in the solar wind pressure initially compresses the magnetosphere and the solar wind's magnetic field interacts with the Earth’s magnetic field and transfers an increased energy into the magnetosphere. Both interactions cause an increase in movement of plasma through the magnetosphere, and an increase in electric current in the magnetosphere and ionosphere [125].

The disturbance in the interplanetary medium which drives the geomagnetic storm may be due to a solar coronal mass ejection (CME) or a high-speed stream (co-rotating interaction region or CIR) of the solar wind originating from a region of weak magnetic field on the Sun's surface. The frequency of geomagnetic storms is increased and decreased with the sunspot cycle. CME driven storms are more common during the maximum of the solar cycle and CIR driven storms are more common during the minimum of the solar cycle [126].

1.14 Investigations on Extrasolar Planets and Artifacts

An extrasolar planet or exoplanet is a planet outside the solar system. Most of the exoplanets are massive giant planets similar to that of Jupiter. Strong Stellar Irradiation flux from the parent star interacts with all relevant depths of the planetary atmosphere with no need for a pre-assumed albedo. Rayleigh scattering increases the extrasolar giant planets flux by orders of magnitude short-ward of the Ca ii H and K doublet (3930A°), and the spectral features of the parent star are exactly reflected. Some kind of extrasolar giant planets of orbit within 0.05 AU of their primary stars, emit detectable infrared radiation [127-129]. This planet has an anomalously large radius, which may be the result of ongoing tidal dissipation, but this explanation requires a non-zero orbital eccentricity (0.03), maintained by interaction with a hypothetical second planet. The large density gradients expected in planet atmospheres can offset the unfavorably large observer lens–to–source lens distance ratio and allow the existence of caustics [130]. Under such conditions of strong lensing, starlight from all points in the planet's shadow is refracted into view, producing a characteristic slowing down of the dimming at ingress [131]. Some of the extra solar planets, which are 60 times larger than earth, are formed by silicon carbide and other carbon compounds. Pulsar planets and low-mass white dwarf planets are especially good candidate members of this new class of planets, but these objects could also conceivably form around stars like the Sun. With the recent discoveries of hundreds of extrasolar planets, the search for planets like Earth and life in the universe is quickly gaining momentum. In the future, large space observatories could directly detect the light scattered from rocky planets, but they would not be able to spatially resolve a planet's surface [132-134].

Interest has developed among researchers, in recent years, to search traces of extraterrestrial intelligence within the solar system. It has already been reported that within the time of existence of our planet, about 10^4 stars capable of having inhabited planets approached the Sun to distances within 1.5 pc which can be covered by space probes using present day technology [135, 136]. Early works [137, 138] have attempted in a scattered way on a search for artifacts which are in orbit, on the Earth or on asteroids. It seems that this list should also include by elaborately considering orbits of different planets and their moons as well [139]. Space activities largely contribute to a long duration pollution of the interstellar medium [140-143]. In a similar way, the interplanetary space of other inhabited planetary systems could contain artifacts. The spontaneous leakage of artifacts into the interstellar medium in absence of interstellar flights is also reasonably expected. This is mainly because of the following three reasons:

(1) A considerable portion of any large artifacts would be ejected by gravitational interaction with the planets. According to computer simulations of the asteroid and comet motion, 10-30 percent of small bodies leave the Solar system [144-146].

(2) Collisions between artifacts or their explosions (like spontaneous explosions of Earth satellites) in the outer parts of the planetary system could accelerate their debris up to hyperbolic velocities.

(3) Light pressure expels micron-sized debris particles (*e.g.* from rocket engines) out of the planetary system. Hence, technical activities even within a planetary system led to a diffusion of artifacts into the interstellar medium. If there are alien artifacts between the stars, some of them are likely to fall down to Earth at some time. So, it is interesting to estimate the frequency of such events.

1.15 Earth-Moon System as a Collector of Alien Artifacts

Three decades of conventional SETI-experiments cover only a negligible part of the Galaxy's age, but more effective approaches could cover billions of years without any speculations about the desire and means of extraterrestrial intelligence (ETI) to communicate with us. It is shown that the Moon is a good indicator of ETI presence in the considerable part of the Galaxy during past 4 Gyr. Moreover, the Earth is a natural collector of ETI artefacts (debris, trash etc.) which could spontaneously fall on our planet [147-150].

The rare oxygen-containing atmosphere of the Earth (a good biosphere indicator) is detectable from interstellar distances. Moreover, ~150 stars of different types have flown through the Solar System at distances $<10^4$ a. u. Hence our unique planet could attract ETI attention. The Moon is a convenient site for long-term observation of the Earth. There is a variety of arguments for placing equipment for prolonged Earth monitoring on the Moon rather than in orbit or on the Earth itself, e.g.

1. Through screening of at least half of the meteorite flow and ionizing radiation, the equipment lifetime on the lunar surface may be at least twice as long as that of a spacecraft in orbit.
2. Orbit stabilization and orientation of an apparatus on the Moon is simple and possible without fuel for jets.
3. Lunar soil can be used to support life of the station personnel, thermal control and repair of equipment.
4. A lunar mission may be effectively concealed from terrestrial aboriginals where as an orbital spacecraft is much easier to reveal.
5. The indisputable advantages of the Moon as an intermediate base for flights to the Earth and planets are clearly demonstrated by the rise of NASA, ESA and NASDA lunar interest [151].

1.16 Interstellar Communications

It was proposed [152] that sky noise windows may be taken as an appropriate alternative search strategy for interstellar communication. For the purpose, the sky noise windows where the hydrogen line and the water line are the two contributors for producing the so called "water holes" can be exploited as an alternative interstellar communication. More specifically, the hydrogen (H) line at 21 cm and the hydroxyl radical (OH) line at 18 cm are combined for producing one water hole between 18 cm and 21 cm while the 2 mm line of start of water molecule extending up to 14 mm line provides a second water hole. Discovered in 1951, the 21 cm hyperfine line of atomic hydrogen has become one of the best-studied spectral features in radioastronomy. Used to trace the distribution and velocity of atomic gas in our own and in other galaxies, the line has made an enormous contribution to our understanding of galactic structure. Now, forty years after the line's discovery, a low-cost receiving system can be assembled from items of consumer electronics to enable study of the line in an undergraduate laboratory. The Drake equation has been utilized to investigate the number of probable civilizations in the Milky Way galaxy by computing the parameters in the equation involved, in a more realistic fashion. The two extreme boundaries have been first determined from the conservative and optimistic point of view and thereby extracting the realistic parameters the probable number of broadcasting civilizations has been estimated [153].

The SETI project has for the past several decades been conducting a search for signals being transmitted by extraterrestrial life located outside the solar system, primarily in the radio frequencies of the electromagnetic spectrum. Special attention has been given to the Water Hole, the frequency of one of neutral hydrogen's absorption lines, due to the low background noise at this frequency and its symbolic association with the basis for what is likely to be the most common system of biochemistry. The regular radio pulses emitted by pulsars were briefly thought to be potential intelligent signals; the first pulsar to be discovered was originally designated "LGM-1", for "Little Green Men." They were quickly determined to be of natural origin, however [154]. Several attempts have been made to transmit signals to other stars as well, see "Realized projects" at Active SETI and Interstellar Radio Messages. One of the earliest and most famous was the 1974 radio message sent from the largest radio telescope in the world, Arecibo, in Puerto Rico. An extremely simple message was aimed at a globular cluster of stars known as M13 in the Milky Way Galaxy and at a distance of 30,000 light years from our solar system. These efforts have been more symbolic than anything else, however. Further, a possible answer needs double the travel time, i.e. tens of years (near stars) or 60,000 years (M13). Sending Interstellar Radio Messages between stars may prove to be optimal for many applications [155].

Data Sources

1. http://www.nrao.edu/whatisra/hist_prehist.shtml
2. http://www.jive.nl/astronomers-demonstrate-global-internet-telescope

References

1. A. B. Bhattacharya, S. Sarkar and R Bhattacharya, "An alternative search strategy for interstellar Communication", *Indian Journal of Physics,* 84 (5), 511-515, 2010.

2. R. A. Howard, J. D. Moses, D. G. Socker, K. P. Dere, J. W. Cook, "Sun Earth Connection Coronal and Heliospheric Investigation (SECCHI)", *Solar Variabilit and Solar Physics Missions Advances in Space Research*, 29, 2017–2026, 2002.

3. A. Driesman, S. Hynes and G. Cancro, "The STEREO Observatory", *Space Science Reviews*, 136, 17–44, 2008.

4. G. Westerhout, "The early history of radio astronomy" *Annals of the New York Academy of Sciences*, (Education in and History of Modern Astronomy), 189, 211-218, 1972.

5. F. Ghigo (ed.). Pre-History of Radio Astronomy, Retrieved 2010-04-09.

6. J. S. Hey. "The Radio Universe", 2nd Ed., Pergamon Press, Oxford-New York, 1975.

7. W. T. Sullivan III, *Classics in Radio Astronomy*. Reidel Publishing Company, Dordrecht, 1982.

8. K. Jansky, "Directional studies of atmospherics at high frequencies", *Proceedings of the Institute of Radio Engineers*, 20, 1920, 1932.

9. P. Robertson, "Beyond Southern Skies: Radio Astronomy and the Parkes Telescope", Cambridge University Press, 1992.

10. R. C. Jennison, *Introduction to Radio Astronomy*, Philosophical Library, New York, 1967.

11. W. T. Sullivan III, *Cosmic Noise: A History of Early Radio Astronomy*, Cambridge University Press, 2009.

12. J. S. Hey, *The Evolution of Radio Astronomy.* Neale Watson Academic, New York, 1973.

13. G. Reber and J. L. Greenstein, "Radio-Frequency Investigations of Astronomical Interest", *Observatory*, 67, 15, 1947.

14. L. L. E. Braes, "Radio Continuum Observations of Stellar Sources", *IAU Symposium No. 60, Maroochydore, Australia, September 3–7, 1973,* 60, 377–381, 1974.

15. D. R. Lorimer, M. Bailes, M. A. McLaughlin, D. J. Narkevic and F. Crawford, "A Bright Millisecond Radio Burst of Extragalactic Origin", *Science*, 318, 777-780, 2007.

16. D. Thornton, et al. "A Population of Fast Radio Bursts at Cosmological Distances", *Science*, 341, 53-56, 2013.

17. S. Burke-Spolaor, M. Bailes, R. Ekers, J. P. Macquart, F. Crawford III, "Radio Bursts with Extragalactic Spectral Characteristics Show Terrestrial Origins", *The Astrophysical Journal*, 727, 5, 2011.

18. P. Saint-Hilaire, A. O. Benz and C. Monstein, "Short-duration Radio Bursts with Apparent Extragalactic Dispersion", *The Astrophysical Journal*, 795, 6, 2014.

19. K. S. Obenberger, G. B. Taylor, J. M. Hartman, J. Dowell, S. W. Ellingson, J. F. Helmboldt, P. A. Henning, M. Kavic, F. K. Schinzel, J. H. Simonetti, K. Stovall and T. L. Wilson, "Detection of Radio Emission from Fireballs" *The Astrophysical Journal Letters*, 788, 6, 2014.

20. K. W. Bannister and G. J Madsen, "A Galactic origin for the fast radio burst FRB010621", *Monthly Notices of the Royal Astronomical Society*, 440, 353-358, 2014.

21. E. F. Keane, B. W. Stappers, M. Kramer and A. G. Lyne, "On the origin of a highly dispersed coherent radio burst", *Monthly Notices of the Royal Astronomical Society: Letters*, 425, L71-L75, 2012.

22. J. H. Oort, "Measures of the 21-cm Line Emitted by Interstellar Hydrogen," *Vistas in Astronomy* (*ELSEVIER*), 1, 607-616, 1955.

23. J. H. Oort, "Radio-frequency Studies of Galactic Structure," *Handbuch der Physik*, 53, 100-28, 1959.

24. G. W. Rougoor and J. H. Oort, "Neutral Hydrogen in the Central Part of the Galactic System," in Paris Symposium on Radio Astronomy, IAU Symposium no. 9 and URSI Symposium no. 1, held 30 July - 6 August 1958, ed. by R.N. Bracewell (Stanford University Press, Stanford, CA, 1959), 416–422.

25. G. W. Rougoor and J. H. Oort, "Distribution and Motion of Interstellar Hydrogen in the Galactic System with Particular Reference to the Region within 3 Kiloparsecs of the Center," *Proceedings of the National Academy of Sciences*, 46, 1-13, 1960.

26. J. H. Oort, "The Formation of Galaxies and the Origin of the High-Velocity Hydrogen," *Astronomy & Astrophysics*, 7, 381-404, 1970.

27. Van de Hulst and C. Hendrik, "Radio Waves from Space: Origin of Radio waves", Ned. tijd. Natuurkunde, 11, 210-221, 1945.

28. I. S. Shklovski, "A Monochromatic Radio Emission from the Galaxy and the Possibility of Observing It", originally published in Russian in *Astronomicheski Zhurnal SSSR,* 26, 10-14, 1948.

29. R. S. Dixon, "A search strategy for finding extraterrestrial radio beacons", Icarus, 20, 187-199, 1973.

30. H. J. Ewen and E. M. Purcell "Observation of a line in the Galactic Radio Spectrum: Radiation from Galactic Hydrogen at 1,420 Mc/s", *Nature*, 168, 356, 1951.

31. H. C. van de Hulst, C.A. Muller, & J. H. Oort, "The spiral structure of the outer part of the Galactic System derived from the hydrogen mission at 21 cm wavelength," *Bulletin of the Astronomical Institutes of the Netherlands*, 12, 117-49, 1954.

32. C. A. Muller and J. H. Oort "Observation of a line in the Galactic Radio Spectrum: The interstellar Hydrogen Line at 1,420 Mc.sec, and an Estimate of Galactic Rotation," *Nature,* 168, 357-358, 1951.

33. C. A. Muller and G. Westerhout, "A catalogue of 21-cm line profiles," *Bulletin of the Astronomical Institutes of the Netherlands*, 13, 151-195, 1957.

34. J. L. Pawsey and R. N. Bracewell, *Radio Astronomy*. Clarendon Press, 361 pages, 1955

35. J. D. Kraus, *Radio Astronomy*, Cygnus Quasar books, USA, 1986.

36. R. S. French, "An Urban Radio Telescope for Jovian and Solar Emissions at 20.1 MHz", *Swinburne Astronomy Online*, HET608, 1-21, 2009.

37. J. Elliott, E. Atwell and B. Whyte, "First stage identification of syntactic elements: An extraterrestrial signal", Proceedings of IAC '2001, The 52nd International Astronautical Congress, October 15, pp IAA01, 2001.

38. J. M. Jenkins and R. D. Laurance, "Detecting reflected light from close in giant planets using spacebased photometers", *The Astrophysical Journal*, 595, 429-445, 2003.

39. S. A. Wouthuysen, "On the excitation mechanism of the 21 cm interstellar hydrogen emission line", *Physica*, 18, 75–76, 1952.

40. C. Heiles and G. T. Wrixon, "Measurements of Galactic 21-cm Hydrogen", *Methods in Experimental Physics*, 12, 58–77, 1976.

41. S. A. Wouthuysen, "On the excitation mechanism of the 21-cm (radio-frequency) interstellar hydrogen emission line", *Astrophysical Journal*, 57, 31-32, 1952.

42. C. M. Hirata, "Wouthuysen-Field coupling strength and application to high-redshift 21-cm radiation", *Monthly Notices of the Royal Astronomical Society*, 367, 259-274, 2006.

43. S. R. Furlanetto, "The global 21-centimeter background from high redshifts", *Monthly Notices of the Royal Astronomical Society*, 371, 867-878, 2006.

44. J. R. Pritchard and A. Loeb, "Evolution of the 21 cm signal throughout cosmic history", *Physical Review D*, 78, 103511, 2008.

45. D. E. Goddard and D. K. Milne, *Parkes. Thirty Years of Radio Astronomy*, Published by CSIRO Publishing, 1994.

46. F. J. Kerr, J. V. Hindman and B. J. Robinson, "Observations of the 21 cm line in the Magellanic Clouds", *Australian Journal of Physics*, 7, 297 – 314, 1954.

47. R. X. McGee, J. D. Murray and J. A. Milton, "A sky survey of neutral hydrogen at *3L* 21 cm. 2. The detailed distribution of

low velocity gas", *Australian Journal of Physics*, 16, 136 -170, 1963.

48. W. N. Christiansen and J. V. Hindman, "A preliminary survey of 1420 Mc/s. line emission from galactic hydrogen", *Australian Journal of Scientific Research*, 5, 437-455, 1952.

49. T. R. Santo and Syed Ashraf Uddin, "Mapping the Spiral Structure of the Milky Way Galaxy at 21cm Wavelength Using the SALSA Radio Telescope of Onsala Space Observatory", *International Journal of Astronomy*, 2, 37-42, 2013.

50. Q. Yin, S. B. Jacobsen and K. Yamashita, "A short time scale for terrestrial planet formation from Hf- W Chronometry of meteorites," *Nature*, 418, 949-951, 2002.

51. S. Joardar, S. Bhattacharyya, A. B. Bhattacharya and C. R. Datta, "Radio Astronomy and Supersynthesis: A Survey," *Progress In Electromagnetics Research* B, 22, 73-102, 2010.

52. J. Kraus, *Antennas*, McGraw-Hill, 361-364, 1950.

53. G. Reber, "A play entitled the beginning of radio astronomy", *Journal of the Royal Astronomical Society of Canada*, 82, 93-106, 1988.

54. R. Haynes, R. Haynes and R. McGee, "Explorers of the Southern Sky: A History of Australian Astronomy", Cambridge University Press, 541 pages, 1996. ISBN 0-521-36575-9,

55. C. M. Matthews, "The Arecibo Ionospheric Observatory", Congressional Research Service, 7-5700, R40437, 2012.

56. A. Verdoni and C. Denker, "The Local Seeing Environment at Big Bear Solar Observatory", *The Publications of the Astronomical Society of the Pacific*, 119, 793-804, 2007.

57. N. D. R. Bhat, J. M. Cordes, P. J. Cox, J. S. Deneva, T. H. Hankins, T. J. W. Lazio and M. A. McLaughlin, "An Arecibo Search for Pulsars and Transient Sources in M33", *The Astrophysical Journal*, 732, 14, 2011.

58. S. Ananthakrishnan, "The Giant Metrewave Radio Telescope", *Journal of Astrophysics and Astronomy I*, 16, 427–435, 1995.

59. C. H. Ishwara-Chandra, A. P. Rao, M. Pandey, R. K. Manchanda, P. Durouchoux, "Low Frequency Radio Observations of GRS1915+105 with GMRT", *Chinese Journal of Astronomy and Astrophysics*, 5, 87–92, 2005.

60. L. Yount, "Modern Astronomy: Expanding the Universe", An imprint of Infobase Publishing, New York, 2006.

61. J. Fuentes-Fernandez, C. E. Parnell and A. W. Hood, "Magneto-hydrodynamics dynamical relaxation of coronal magnetic fields: II. 2D Magnetic X-Points" *Astronomy and Astrophysics*. 536, 32, 2011.

62. P. Foukal, C. Fröhlich, H. Spruit and T. M. L. Wigley, "Variations in solar luminosity and their effect on the Earth's climate", *Nature* 443, 161-166, 2006.

63. R. A. Harrison, A. W. Hood and C. D. Pike, "*Off-Limb EUV line profiles and the search for wave activity in the low corona*" *Astronomy and Astrophysics*, 392, 319-327, 2002.

64. B. S. Rathore, S. C. Kaushik, R. S. Bhadoria, K. K. Parashar and D. C. Gupta, "Sunspots and geomagnetic storms during solar cycle-23" *Indian Journal of Physics*, 86, 563-567, 2012.

65. N. Kumar, P. Kumar, A. Kumar and R. Chauhan, "Role of equilibrium plasma flow on damping of slow MHD waves", *Indian Journal of Physics*. 85, 1879, 2011.

66. J. L. Pawsey and R. N. Bracewell, *Radio Astronomy*. Clarendon Press, 361, 1955.

67. G. A. Dulk, Y. Leblanc, R. J. Sault, and S. J. Bolton, "Jupiter's magnetic field as revealed by the synchrotron radiation belts II. change of the 2-D brightness distribution with DE", *Astronomy and Astrophysics*, 347, 1039-1045, 1999.

68. G. A. Dulk, Y. Leblanc, R. J. Sault, S. J. Bolton, J. H. Waite and J. E. P. Connerney, "Jupiter's magnetic field as revealed by the synchrotron radiation belts I. Comparison of a 3-D reconstruction with models of the field", *Astronomy and Astrophysics*, 347, 1029-1038, 1999.

69. J. H. Rogers, *The Giant Planet Jupiter*, Cambridge, Cambridge University Press, 1995.

70. A. B. Bhattacharya, S. Joardar, R. Bhattacharya, A. Bhoumick, A. Nag, M. Debnath and D. Halder, "Reception of Jovian radio signals in a tropical station", *International Journal of Engineering Science and Technology*, 2, 5704-5713, 2010.

71. P. C. W. Fung, "Excitation of cyclotron radiation in the forward subluminous mode and its application to Jupiter's decametric emissions", *Elsevier*, 14, 469–481, 1966.

72. J. D. Kraus, "Impulsive Radio Signals from the Planet Venus", *Nature*, 178, 33, 1956.

73. J. D. Kraus, "Class II Radio Signals from Venus at a Wave-length of II Metres", *Nature,* 178, 159-160, 1956.

74. F. E. Ross, "Photographs of Venus", *Astrophysical Journal*, 68, 57, 1928.

75. W. M. Sinton and J. Strong, "Radiometric Observations of Venus", *Astrophysical Journal*, 131, 470, 1960.

76. J. D. Kraus, "Rotation Period of the Planet Venus as determined by Radio Observations", *Nature*, 178, 687-688, 1956.

77. Mona Lisa, A. A. Khwaja, M. Q. Jan, "Seismic Hazard Assessment of the NW Himalayan Fold-and-Thrust Belt, Pakistan, Using Probabilistic Approach", *Journal of Earthquake Engineering*, 11, 257- 301, 2007.

78. D. Shanker, R. B. S. Yadav, H. N. Singh, "On the seismic risk in the Hindukush-Pamir-Himalaya and their vicinity", *Current Science*, 92, 1625-1630, 2007.

79. Q. Yin, S. B. Jacobsen and K. Yamashita, "A short time scale for terrestrial planet formation from Hf- W Chronometry of meteorites," *Nature*, 418, 949-951, 2002.

80. R. Arlt, J. Rendtel, P. Brown, V. Velkov, W. K. Hocking and J. Jones, "The 1998 outburst and history of the June BooÈtid meteor shower", *Monthly Notices of the Royal Astronomical Society*, 308, 887-896, 1999.

81. B. Saha and A. K. Sen, "Long period fading of LF radio signals and geomagnetic activity during major meteor showers," *Annales de Geophysique*, 36, 61-66, 1980.

82. P. Jenniskens, P. S. Gural, L. Dynneson, B. J. Grigsby, K. E. Newman, M. Borden, M. Koop and D. Holman, "CAMS: Cameras for Allsky Meteor Surveillance to establish minor meteor showers", Icarus, 216, 40–61, 2011.

83. S. P. Gupta, "Ionisation layers over the magnetic equator during meteor shower days," *Advances in Space Research*, 10, 105–108, 1990.

84. B. Saha and A. K. Sen, "Attenuation of LF radio signals due to meteor trains," Annales de Geophysique, 33, 503-511, 1977.

85. A. K. Sen and B. Saha, "Radio wave propagation and ionization of meteoric constituents", *Nature*, 255, 313-316, 1975.

86. J. D. Kraus, "Relation of 11-Metre Solar System Phenomena to Solar Disturbances", *Nature*, 179, 371, 1957.

87. J. L. Pawsey, "Solar radio frequency radiation", *Proceedings of the Institution of Electrical Engineers*, 97, 290-310, 1950.

88. M. Stix, *The Sun: an Introduction*. Springer, 2004.

89. E. F. Keane, B. W. Stappers, M. Kramer and A. G. Lyne, "On the origin of a highly dispersed coherent radio burst", *Monthly Notices of the Royal Astronomical Society: Letters*, 425, L71-L75, 2012.

90. C. A. Katz, J. N. Hewitt, B. E. Corey and C. B. Moore, "A Survey for Transient Astronomical Radio Emission at 611 MHz" *The Publications of the Astronomical Society of the Pacific*, 115, 675-687, 2003.

91. M. J. Aschwanden, H. J. Wiehl, A. O. Benz and S. R. Kane, "Correlation of solar decimetric radio bursts with X-ray flares", *Solar Physics*, 97, 159-172, 1985.

92. W. N. Christiansen and J. A. Warburton, "The distribution of radio brightness over the solar disk at a wavelength of 21 cm. Part III. The Quiet Sun two-dimensional observations", *Australian Journal of Physics*, 8, 474-486, 1955.

93. R. Behlke, "Solar radio bursts and low frequency radio emissions from space" IRF Scientific Report 275, Swedish Institute of Space Physics, 2001.

94. C. Caroubalos, et. al. "ARTEMIS IV Radio Observations of the 14 July 2000 Large Solar Event", Solar Physics, 204, 165-177, 2001.

95. Y. Liu, J. G. Luhmann, S. D. Bale, and R. P. Lin "Relationship Between a Coronal Mass Ejection-Driven Shock

And A Coronal Metric Type II Burst" *The Astrophysical Journal*, 691, L151–L155, 2009.

96. C. Caroubalos et. al. "Joined observations of solar radio bursts and cosmic ray events recorded by ARTEMIS IV radio spectrograph and Athens Cosmic Ray Station (Super 6NM-64) of the University of Athens" *Hellenic Astronomical Society Conference* 20-22 September Crete, Greece 2001.

97. A. B. Bhattacharya, S. Joardar, and R. Bhattacharya, *Astronomy and Astrophysics*, Infinity Science Press, Massachusetts, USA 2008.

98. G. Li and G. P. Zank, "Mixed particle acceleration at CME-driven shocks and flares", *Geophysical Research Letters*, 32, L02101, 2005.

99. G. A. Dulk, "Radio-Emission from the Sun and Stars", *Annual Review of Astronomy and Astrophysics*, 23, 169-224, 1985.

100. V. L. Verschuur and K. Kellermann, *Galactic and Extragalactic Radio Astronomy*, Springer-Verlag, New York, second edition, 1988,

101. J. D. Richardson, C. Wang and K. I. Paularena "The solar wind: from solar minimum to solar maximum". *Advances in Space Research*, 27, 471-479, 2001.

102. K. Kudela, J. Rybak, A. Anatalova and M. Storini, “Time evolution of low-frequency periodicities in cosmic ray intensity”, *Solar Physics*, 2005, 165-175, 2002.

103. K. Labitzke and K. Matthes “Eleven-year solar cycle variations in the atmosphere: observations, mechanisms and models”. *Holocene*, 13, 311-317, 2003.

104. D. Shindell, D. Rind, N. Balachandran, J. Lean and P. Lonergan, “Solar cycle variability, ozone and climate”. *Science*, 284, 305-308, 1999.

105. F. Boberg and H. Lundstedt, “Solar wind variations related to fluctuations of the North Atlantic Oscillation”, *Geophysical Research Letters*, 29, 1718, 2002.

106. E. Bard, G. Raisbeck, F. Yiou and J. Jouzel, “Solar irradiance during the last 1200 years based on cosmogenic nuclides”, *Tellus*, 52B, 985-992, 2000.

107. E. W. Cliver, V. Boriakoff and J. Feynman, “Solar variability and climate changes: geomagnetic and aa index and global surface temperature”, *Geophysical Research Letters*, 25, 1035-1038, 1998.

108. D. Alexander, I. G. Richardson and T. H. Zurbuchen, "A Brief History of CME Science", *Space Science Reviews*, 123, 3–11, 2006.

109. T. Amari, J. F. Luciani, J. J. Aly and M. Tagger, "Plasmoid formation in a single sheared arcade and application to coronal mass ejections", *Astronomy and Astrophysics*, 306, 913, 1996.

110. T. Amari, J. F. Luciani, Z. Mikic and J. Linker, "A Twisted Flux Rope Model for Coronal Mass Ejections and Two-Ribbon Flares", *Astrophysical Journal Letter*, 529, L49–L52, 2000.

111. T. Amari, J. F. Luciani, J. J. Aly, Z. Miki/c and J. Linker, "Coronal Mass Ejection: Initiation, Magnetic Helicity, and Flux Ropes. I. Boundary Motion-driven Evolution", *Astrophysical Journal*, 585, 1073–1086, 2003a.

112. T. Amari, J. F. Luciani, J. J. Aly, Z. Miki/c and J. Linker, "Coronal Mass Ejection: Initiation Magnetic Helicity, and Flux Ropes. II. Turbulent Diffusion-driven Evolution", *Astrophysical Journal,* 595, 1231–1250, 2003b.

113. T. Amari, J. F. Luciani and J. J. Aly, "Coronal Magnetohydrodynamic Evolution Driven by Sub-photospheric Conditions", *Astrophysical Journal Letter*, 615, L165–L168, 2004.

114. M.D. Andrews, "The Front-to-Back Asymmetry of Coronal Emission", *Solar Physics*, 208, 317–324, 2002.

115. S. K. Antiochos, C. R. DeVore and J. A. Klimchuk, "A Model for Solar Coronal Mass Ejections", *Astrophysical Journal*, 510, 485–493, 1999.

116. V. Archontis and A. W. Hood, "A Flux Emergence Model for Solar Eruptions", *Astrophysical Journal Letter*, 674, L113–L116, 2008.

117. T. Henke, J. Woch, U. Mall, S. Livi, B. Wilken and R. Schwenn, "Differences in the O_7^+/O_6^+ ratio of magnetic cloud and non-cloud coronal mass ejections", *Geophysical Research Letters*, 25, 3465-3468, 1998.

118. C. Wang, J. D. Richardson and K. I. Paularena, "Predicted Voyager observations of the Bastille Day 2000 coronal mass ejection", *Journal of Geophysical Research,* 106, 13007-13013, 2001.

119. Y. D. Liu, J. D. Richardson, C. Wang and J. G. Luhmann. "Propagation of the 2012 March coronal mass ejections from the sun to heliopause", *The Astrophysical Journal Letters*, 788, L28, 2014.

120. Y. Liu, J. D. Richardson, J. W. Belcher, J. C. Kasper and H. A. Elliott, "Thermodynamic Structure of Collision-dominated Expanding Plasma: Heating of interplanetary Coronal Mass

Ejections", *Journal of Geophysical Research,* 111, A01102, 2006.

121. C. Wang and J. D. Richardson, Interplanetary Coronal Mass Ejections Observed by Voyager 2 between 1 and 30 AU, *Journal of Geophysical Research*, 109, A06104, 2004.

122. Wang, C., J. D. Richardson, and L. Burlaga, Propagation of the Bastille Day 2000 CME shock in the outer heliosphere, *Solar Physics,* 204, 413-423, 2001.

123. H. Maeda, "Variation in Geomagnetic Field", *Space Science Reviews*, 8, 555-590, 1968.

124. G. Richardson, E. W. Cliver and H. V. Cane, "Sources of geomagnetic activity over the solar cycle: relation importance of coronal mass ejections, high speed streams and slow solar wind", *Journal of Geophysical Research*, 105, 18203-18214, 2000.

125. S. K. Antiochos, R. B. Dahlburg and J. A. Klimchuk, "The magnetic field of solar prominences", *The Astrophysical Journal Letters*, 420, L41–L44, 1994.

126. D. Charbonneau, L. Allen, S. Megeath, G. Torres, R. Alonso, T. Brown, R. Gilliland, D. Latham, G. Mandushev, F. O'Donovan and A. Sozzetti, "Detection of Thermal Emission from an Extrasolar Planet", *The Astrophysical Journal,* 626, 523–529. 2005.

127. D. Deming, S. Seager, J. Richardson and J. Harrington, "Infrared Radiation from an Extrasolar Planet", *Nature,* 434 740–743. 2005.

128. L. V. Ksanfomality "Transits of extrasolar planets", *Solar System Research*, 41, 463-482, 2007.

129. S. Seager and D. D. Sasselov "Extrasolar Giant Planets under Strong Stellar Irradiation", *The Astrophysical Journal*, 502, L157–L161, 1998.

130. S. Seager and Lam Hui "Constraining the rotation rate of transiting extrasolar planets by oblateness measurements", *The Astrophysical Journal*, 574, 1004–1010, 2002.

131. K. V. Kholshevnikov, E. D. Kuznetsov "Selection Effect in Semimajor Axes of Orbits of Extrasolar Planets", *Solar System Research*, 36, 466-477, 2002.

132. L. L. Sokolov "Orbital Evolution of Extrasolar Planets", *Solar System Research*, 36, 403-408, 2002.

133. B. Lopez and R. G. Petrov "Direct Detection of Hot Extrasolar Planets Using Differential Interferometry", *ESO Astrophysics Symposia: Extrasolar Planets to Cosmology*, 565-570, 2000.

134. A. R. Martin, Project Daedalus: bombardment by interstellar material and its effects on the vehicle. *Journal of the British Interplanetary Society, supplement*, S116-S121, 1978.

135. H. Heusmann, “Space Debris”, *Spaceflight*, *35*, 182-184, 1993.

136. M. Duncan, T. Quinn and S. Tremaine, “The formation and extent of the solar system cloud”, *Astronomical Journal*, 94, 1330-1338, 1987.

137. G. V. Foster, “Non-Human Artefacts in the Solar System”, *Spaceflight*, 14, 447-453, 1972.

138. A. V. Arkhipov, “Astro-dynamical aspect of paleohistology”, *Research Institute on Anomalous Phenomena (RIAP) Bulletin*, 1, 7-10, 1994.

139. A. V. Arkhipov, “Extraterrestrial Artefact?”, *Spaceflight*, 37, 94, 1995.

140. P. Farinella, Ch. Froeschlé, C. Froeschlé, R. Gonczi, G. Hahn, A. Morbidelli, G. B. Valsecchi, “Asteroids failing into the Sun”, *Nature*, 371, 314-317, 1994.

141. S. I. Ipatov, "Migration of Small Bodies to the Earth", *Astronomicheskii Vestnik*, 29, 304-330, 1995.

142. C. W. Allen, *Astrophysical Quantities*, The Athlone Press, London, 1973.

143. A. V. Tutukov, "Search of Planets Around Low Mass Main Sequence Stars", *Astron. Zhur.* 72, 400-409, 1995.

144. M. D. Papagiannis, "Natural Selection of Stellar Civilizations by the Limits of Growth", *Quarterly Journal of the Royal Astronomical Society*, 25, 309-318, 1984.

145. V. A. Bronshten, *Physics of Meteoric Phenomena*, Geophysics and Astrophysics Monographs, Nauka, Moscow, 1981.

146. J. Heidmann, "SETI-Programmes All Over the World (and further out)", *Journal of the British Interplanetary Society,* 48, 447-452, 1995.

147. G. V. Foster, "Non-Human Artefacts in the Solar System", *Spaceflight*, 14, 447-453, 1972.

148. A. V. Arkhipov, "On the Importance of Nonclassical SETI", *The Observatory*, 113, 306-307, 1993.

149. Lager, M. Pirre -and F. J. Marceau, "Relevance of Oxygen and Ozone Detections in the Search for Primitive Life in Extra Solar Planets", *Advances in Space Research*, 14, 117-122, 1994.

150. V. Arkhipov, "Archaeological Elements of Lunar Exploration", *Solar System Research*, 28, 467-469, 1994.

151. A. B. Bhattacharya, S. S. Banerjee and R. Bhattacharya "A Roadmap for detecting Extraterrestrial Intelligent Life", PIERS Proceedings, Beijing, China, March 2327, 1400-1403. 2009.

152. M. M. Ross "The Drake Equation and habitable planets", *The Search for Extraterrestrials*, Springer Praxis Books, 41-55, 2009.

153. A. L. Zaitsev, "Sending and Searching for Interstellar Messages". *Acta Astronautica,* 63, 614–617, 2008.

154. R. A. Freitas, Jr. "Interstellar Probes: A New Approach to Seti", *Journal of the British Interplanetary Society*, 33, 95–100, 1980.

Chapter 2

TECHNIQUES & INSTRUMENTATIONS

2.1 Established Methods for Detecting Extrasolar Planets

In recent years scientists have realized that intelligent life may well be found throughout the universe and we are, perhaps, not the only civilization in our galaxy. The work suggests an alternative search strategy through sky noise windows emphasizing both the "water holes" and the associated wavelengths. It also suggests that instead of pointing the antenna at known stars, an all-sky search would be more appropriate for interstellar communication [1-3]. There are various methods to find the extrasolar planets. Out of many attempts so far, the following indirect methods have revealed successful results. Established detection methods are:

(i) Astrometric Method,

(ii) Radial velocity methods and Doppler shift measurement,

(iii) Dynamical perturbations,

(iv) Gravitational Microlensing,

(v) Transit Method,

(vi) Pulsar Timing Method,

(vii) Circumstellar Disks Method,

(viii) Direct Imaging Method,

(ix) Observations from Space and

(x) MOP and Phoenix.

These methods are discussed here [4-6].

2.1.1 Astrometric Method

Astrometry is the method of measuring the position of a star in the sky and precisely observing how that position changes with time. One great advantage of this method is that it is highly sensitive to planets with large orbits. This makes astrometric method complementary to other methods that are most sensitive to planets with small orbits. However, long observation times required – may be years and even possibly decades [4, 7].

2.1.2 Radial Velocity Methods and Doppler Shift Measurement

The observed frequency of radiation emitted by a source (ν_{obs}) moving at a constant velocity is different from that of a source stationary with respect to the observer (ν_{source}). The relative frequencies, according to relativistic expression is given by the relation [8],

$$\frac{\nu_{obs}}{\nu_{source}} = \sqrt{\frac{1-\frac{v}{c}}{1+\frac{v}{c}}} \qquad (2.1)$$

When $v \ll c$ the frequency shift $\Delta\nu = \nu_{obs} - \nu_{source}$ can be expressed as,

$$\frac{\Delta\nu}{\nu_{source}} \approx -\frac{v}{c} \qquad (2.2)$$

More generally, for motion in an arbitrary direction the non-relativistic expression becomes,

$$\frac{\Delta\nu}{\nu_{source}} \approx -\frac{v\cos\theta}{c} \qquad (2.3)$$

where, θ is the angle between velocity vector and the line of sight to the observer. This is schematically shown in Figure 2.1.

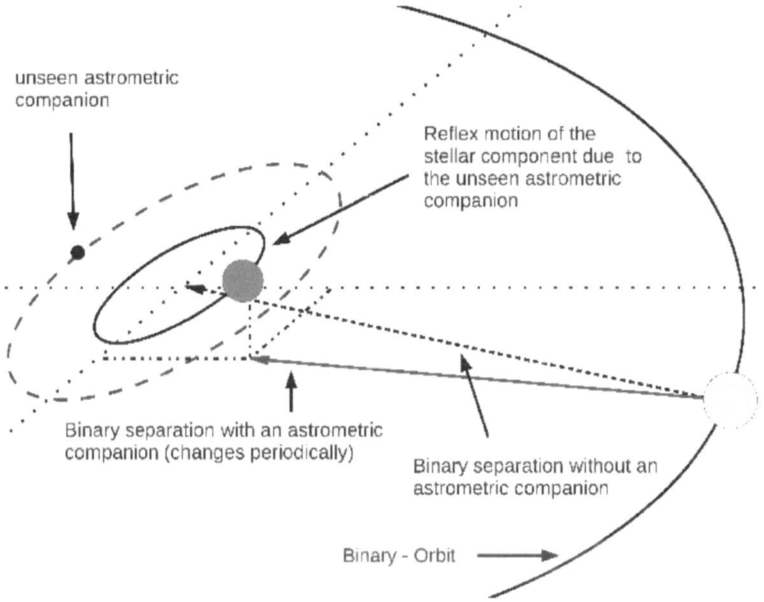

Figure 2.1 A schematic illustration of the spectroscopic measurement of stellar wobble due to the presence of a planet

For a circular orbit coplanar with the observer's line of sight then the maximum relative velocity is $v = v_{*}$, depending on whether the star is coming towards the observer (frequency increase, blue shift) or is going away from the observer (frequency decrease, red shift).

2.1.3 Dynamical Perturbations

Let us consider, two objects in orbit about each other; one is star and the other one is a planets of mass M_{*} and M_p respectively. Let R_* and let R_p be their radii with respect to centre of mass of the system. In general, the orbital velocity of the star is given by

$$v_*^2 = \frac{G(M_* + M_p)R_*^2}{R_p^3} \qquad (2.4)$$

Considering center of mass coordinate system,

we have $M_* R_* = M_p R_p$ and assuming that $M_p \ll M_*$

Thus, we obtained,

$$v_*^2 = \frac{GM_p^2}{R_p M_*} \qquad (2.5)$$

This is schematically shown in the Figure 2.2.

2.1.4 Gravitational Microlensing

Gravitational microlensing happens in principle when the gravitational field of a star acts like a lens, magnifying the light of a distant background star. The effect occurs only when the two stars are almost aligned. Lensing events are brief, lasting for weeks or days, as the two stars and the Earth are all moving related to each other. Gravitational lensing is the achromatic deflection of light due to gravity. A basic schematic of what happen is shown in the Figure 2.3. In this example the resultant image of the source is deflected by an angle α.

$$\alpha = \frac{4GM}{c^2 R} \qquad (2.6)$$

R is the impact parameter, $2GM/c^2$ is the Schwarzschild radius

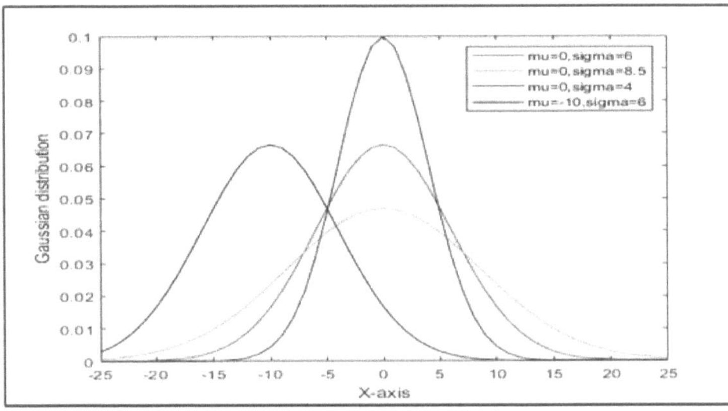

(**a**) two-dimensional Gaussian distribution

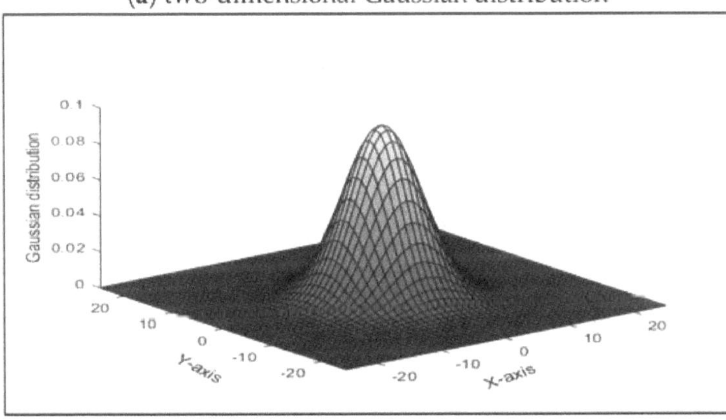

(**b**) Three-dimensional Gaussian distribution

Figure 2.2 2d and 3d discussion of dynamical perturbation, assuming M_* greater than M_p [8]

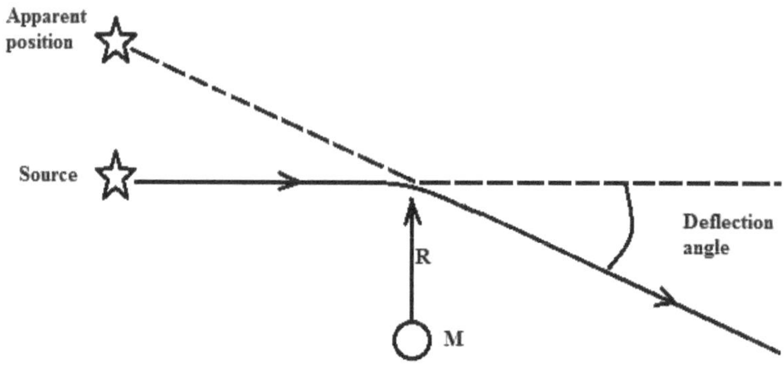

Figure 2.3 An achromatic deflection of light by a mass M between the source and the observer; the fashion analogous to classical optics, the mass M acts like a lens [8]

2.1.5 Transit Method

If a planet passes between the observer and its parent star then the light from the star will be attenuated as the planet passes across the stellar disk. Typically, the attenuation will be significantly less than some 1-2% of the total starlight. The attenuation is the ratio of the planet and star visual disk areas,

$$\frac{\Delta L}{L_*} \simeq \left(\frac{R_P}{R_*}\right)^2 \qquad (2.7)$$

The basic model for a planet transiting in front of its parent star is shown in Figure 2.4. In the upper plot the planet (*P*) can be considered to be moving along a linear path at a stellar latitude δ. As it transits a fraction of the stellar light is blocked and the net luminosity of the system appears to drop. This is shown in the lower plot. The shape of the transit curve in this case is just due to geometric superposition of two uniform circular disks. The luminosity L_* is reduced by ΔL during full transit [8, 9].

The transit duration is given by

$$\tau = \frac{P}{\pi}\left(\frac{R_* cos\delta + R_p}{a}\right) \qquad (2.8)$$

Where P is the orbital period, a is the planet-star separation at transit and δ is the mean stellar latitude of the transit. Now, the duration τ can be directly measured, and the interval between transits also yields P to high precision. The stellar radius R_* can be estimated from the knowledge of stellar structure and the spectroscopy measurement of the star. a can be determined from the Kepler's 3rd law. R_p can be measured from the attenuation $\Delta L/L_*$, and then the latitude δ of the transit can be estimated.

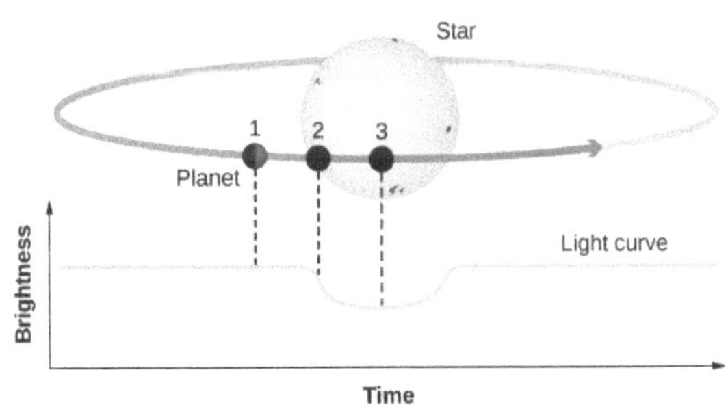

Figure 2.4 The basic model for a planet transiting in front of its parent star

2.1.6 Pulsar Timing Method

A pulsar is a small, ultra dense remnant of neutron star. It is a star that has exploded as a supernova. Pulsars emit radio waves regularly as they rotate. As the intrinsic rotation of a pulsar is regular, slight anomalies in the timing of its observed radio pulses can be used for tracking its motion [5].

2.1.7 Circumstellar Disks Method

Disks of space dust (called debris disks), surround many stars. This dust can be detected as it absorbs ordinary starlight and re-emits it as infrared radiation. Even if the dust particles have a total mass sufficiently less than that of the earth, they can still have a considerably large total surface area that they outshine their parent star in infrared wavelengths. The Hubble Space Telescope is capable of observing dust disks with the help of NICMOS (Near Infrared Camera and Multi Object Spectrometer) instrument [10].

2.1.8 Direct Imaging Method

Current Telescopes are capable of directly imaging planets, e.g. the Gemini Telescope, the Subaru Telescope and the VLT etc. Direct imaging is possible when the planet is especially large (considerably larger than the Jupiter) and widely separated from its parent star. A group of astronomers used the European Southern Observatory's very large telescope array in 2004 and 2005 and also in 2007 to produce images of exoplanets.

Future Detection Methods are:

(i) Observations from space,

(ii) Eclipsing binary minima timing,
(iii) Orbital phase reflected light variations and
(iv) Polarimetry.

2.1.9 Observations from Space

Astronomical measurements from space can be more sensitive than measurements done from the ground, as the distorting effect of the Earth's atmosphere is removed and the instruments can view in infrared wavelengths that do not penetrate the atmosphere. These space probes even have the capability of detecting planets similar to our Earth [4, 11-16]. NASA has cut funding for the Terrestrial Planet Finder in 2007 and the funding has gone towards the Kepler Mission. This is mainly because the transit method will be used by the Kepler Mission to scan a hundred thousand stars simultaneously and thus will be able to collect statistics on the numbers of planets around sun like stars [17].

2.1.10 MOP and Phoenix

The main objectives of the Microwave Observing Program [11, 18-19] abbreviated as MOP, were the following: (i) "Targeted Search" Program, (ii) 800 Specific nearby Stars & General Sky Survey, (iii) Radio Dishes associated with NASA and (iv) A 43-meter dish at Green Bank. Signals were to be analyzed by Spectrum Analyzers each with a capacity of 15 million Channels.

On the other hand, Project "Phoenix" (1995) was largely engaged in:
(i) "Targeted Search" Program,

(ii) Studying about 1000 nearby Sun-like Stars,
(iii) Phoenix conducted observation at the 64-meter Parks Radio Telescope in Australia and
(iv) A 140-ft Telescope of the National Radio Astronomy Observatory in West Virginia, USA.

The project observed 800 Stars over the available channel in the frequency range from 1200 to 3000 MHz.

2.2 Recording Techniques

In our laboratory there are some radio receivers, which are 1420 MHz Hydrogen line Spctra-Cyber receiver, 406.7 MHz Solar Radio telescope, Log Periodic Dipole Array Antenna operates on 50 MHz to 300 MHz, 40 kHz VLF atmospheric Receiver and 92 MHz VHF receiver.

2.2.1 1420 MHz Hydrogen Line Spectra-Cyber

(i) Description of the Radio Telescope:

The Radio Telescope is comprised of two units; one called the front end the other the backend. The front end contains the low noise amplifier and the Receiver Backend contains the converter and signal processing circuitry. Both are powered by +12 Vdc, which is supplied by the backend and the external power supply. This low noise system is constructed with the latest, microwave component technology and, is state of the art for amateur radio astronomy. The 1420 MHz converter is a dual conversion unit, which converts the 1420 MHz. hydrogen region down to the 70MHz. IF frequency. The backend contains a high gain IF amplifier, square law diode detector, computer-controlled DC amplifier, programmable integration control, adjustable offset, and a 12-bit A/D converter, power supply, all controlled by a Basic Stamp II microcomputer! Interface to the outside world is by RS-232 link to an IBM compatible computer on com port #1 or #2. Presently, no other com ports are supported [20-22].

(ii) Single Antenna Hookup:

The feed horn should be mounted at the focus of the dish in whatever manner you choose (The focal point of your antenna should be at the mouth of the horn, not inside at the monopole). The LNA should be connected to the feed horn. A weather proof box should be used to house the front-end module (LNA) [23-25].

(iii) Interferometer Hookup (Two Antennas):

For two antenna systems, two feed-horns are supplied as well as two LNA's. Mounting of the feed-horns and placement of the LNA's are the same as above. A weather proof box should be used to house the front-end module (LNAs). NOTE: Suggested antenna placement should not be more than 25 - 50 feet apart. If more distance is required, a pair of in-line booster amplifiers may be required. The separate antenna cables are connected via a power combiner which also routes the combined signal to the receiver backend. It cannot be stressed enough that any outside modules (LNA), which comprises part of the radio telescope front end, are not fully weather proof! The low noise amplifier/s must be housed in an additional weather proof box, (such as Rubbermaid container) simply painting will not keep the elements out. Damaged electronics may result if these instructions are not followed! Power for the LNA is obtained from the backend via the power cable. The RF output of the LNA is connected to the coax cable and run into the observation area, to the backend unit. The backend unit is meant to be kept in a shirt sleeve environment such as, a habitable room. This unit is not resistant to the elements and must be kept free of temperature extremes and high humidity. Keep in mind that this unit will be connected to a computer (user supplied), which is also very sensitive to environmental conditions [26].

(iv) Instrument Employed for the Present Study:

The Radio Telescope we have used is composed of a front-end unit which includes the low noise preamplifier and a cylindrical feed horn for 1420 MHz. It is constructed and developed by Radio Astronomical Supplies (RAS), USA. The frontend is connected to the backend rack unit by user supplied low loss coaxial cable with type "N" connectors. In addition, 18 Gauge, four conductor (shielded) power wire is used for external LNA power. We implemented a "very low loss" cable LMR-400 flexible cable to minimize a loss of 6 dB or less at 1420 MHz. The arrangement is shown in Figure 2.5.

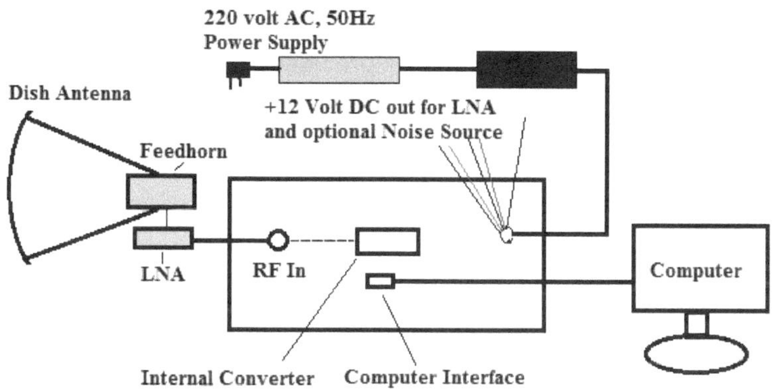

Figure 2.5 Setup diagram for the Spectra-Cyber used for recording 1420 MHz radio signal

Depending on the length of cable run, it was necessary to add an additional low noise preamp in the line. Extremely low loss hard line allowed running over 100 feet with no additional preamplifier. The feed line carries the low noise amplified signals to the telescope backend. The back end is powered by a 220 V AC supply which produces a nominal +/- 12.5 V DC as well as a rear panel output to power the preamplifier and an optional noise source. The instrument communicates with an IBM compatible CPU via an RS-232 dB9 connector. The 1420 MHz signal from the frontend (LNA) enters the rear panel (Receiver Backend) and is fed to a 1420 to 70 MHz dual conversion (Internal) down converter. This converter has 8 MHz bandwidth with the Hydrogen rest frequency at 70.0 MHz. This 8 MHz wide IF signal is passed through a programmable gain IF amplifier and then split between the continuums square law detector and the spectrometer third conversion mixer. The programmable gain IF amplifier is used to compensate for feed line losses and to place the signal in optimum range for the square law detectors [26-27].

The software has been designed and tested on XP and a SVGA (800 x 600) monitor. The software provides a control and hardware setting interface to the spectrometer. The software includes features to enable the user for playing data movies of hundreds of scans at a time, select time coordinate systems for the data files, set the computer clock to within 500 milliseconds of WWV via the internet, and change graphical features such as background or the amplitude axis on the data plot. Final modifications include a frequency shifting for equipment that can shift frequencies around 1420 MHz to a 70 MHz IF, displayed on the X axis. In practice, we can run a single scan and save it to produce the bmp. The 13k bmp can be converted to a 6k gif file with graphic software package. This bmp or gif graphics file can be used for data plots [29].

2.2.2 406.7 MHz Solar Radio Telescope

The radio telescope used for recording the data is basically comprised of two units, one called the front end and the other the backend. The front end contains the low amplifier and the 406.7 MHz internal converter. Both are powered by +12 V dc, which is supplied by the backend and the external power supply. This low noise system is constructed with the latest microwave component technology. The 406.7 MHz converter is a dual conversion unit, which converts the 406.7 MHz down to 70 MHz IF frequencies [29-34]. The block diagram of the experimental setup including the dipole antenna mounting system is exhibited in Figure 2.6.

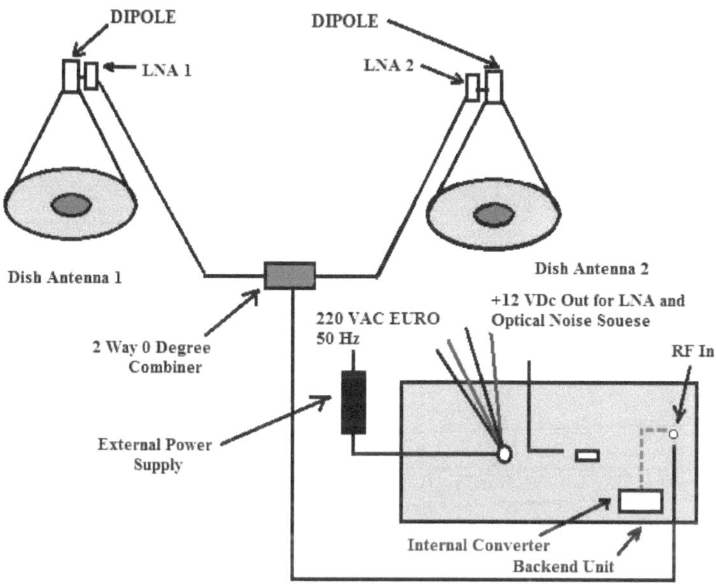

Figure 2.6 Block diagram of the experimental setup including the dipole antenna

2.2.3 Log Periodic Dipole Array Antenna

The log-periodic dipole array (LPDA), is a multi-element narrow-beam antenna, which is used conveniently for the reception of solar burst. The antenna array has impedance and radiation characteristics that are periodic as a logarithmic function of the excitation frequency. It can accommodate a suitable range of frequencies with moderate gain and directionality. Our receiving systems linked to two LPDAs successfully to receive solar signal. Here we use two low noise amplifiers (LNA) to reduce the local noises significantly and thereafter the signal is fed to high frequency amplifier (VHF) to amplify the received signal [35-39]. A schematic diagram of the whole arrangement is shown in Figure 2.7.

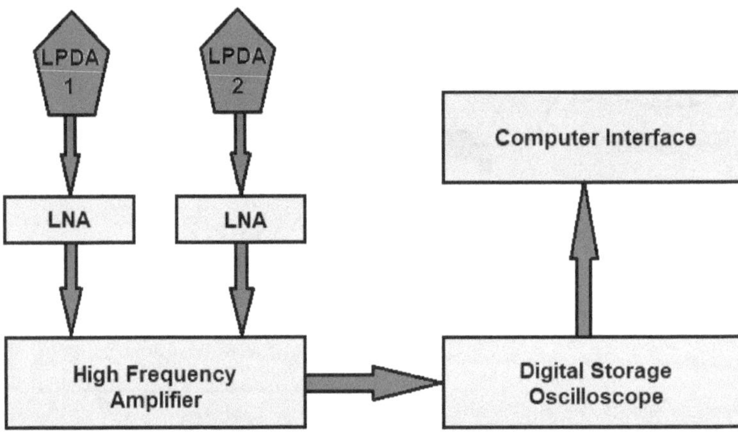

Figure 2.7 Block diagram of Log Periodic Dipole Array

2.2.4 40 kHz VLF atmospheric Receiver

The receiver employed for the observations at 40 kHz is an imported one from Radio Astronomical Supplies (RAS), USA. It is designed in such a manner so that it can successfully handle variation of field intensities due to atmospherics originating due to meteorological disturbances. The unit has tuned r.f. stages followed by a detector and D.C. amplifier which in turn feed the recorder where a master computer is used for recording the atmospherics data in digital form. The type of antenna used with the receiver is governed by the nature of polarization of the signal to be received and also occasionally by the direction of arrival of the wave while the dimension of the antenna with horizontal part 30 ft and vertical part 10 ft largely depends on the wavelengths, the ratio of the signal picked up by the antenna to the inherent receiver noise. The polarization of the atmospheric field is mainly vertical in the designated frequency range. In our observation we used an inverted-L antenna whose vertical part would receive this field with an omni directional azimuthal pattern. The function of the horizontal part is simply to add to the top capacitance of the antenna causing an increase of current to the vertical part. In order to further mention the technical part of the instrument it may be pointed out that the receiver is composed of a very sharp double tuned filter and three stages of amplification in cascade. The gain of the cascade stages is set by a single potentiometer. The output of the amplifiers is fed to a diode detector and then to a 10 second integrator. Figure 2.8 depict the block Diagram of VLF Receiver [29, 43-45].

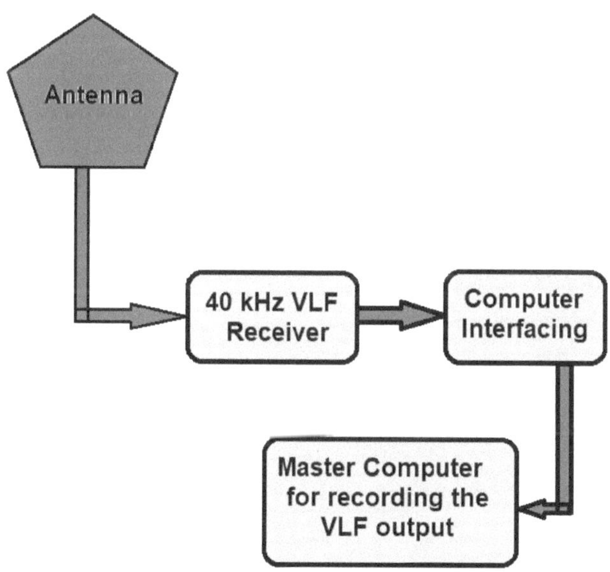

Figure 2.8 Block Diagram of VLF Receiver

2.2.5 92.7 MHz VHF Receiver

We set our experiments to study the propagation effect of radio signal at VHF transmitted at 92.7 MHz from Kolkata. A super heterodyne receiver used to receive this transmitted signal at Kalyani round-the-clock. Measurements have been made by recording audio frequency output of the super heterodyne receiver. In the experimental arrangement we designed the circuit in such a manner that after receiving the electromagnetic waves transmitted by the antenna, the identical voltage is transferred to the receiver input which is amplified by the circuits. The carrier frequency (f_c) and the frequency of the local oscillator (f_0) are fed to the mixer whose output provides the intermediate frequency $(f_i = f_c - f_0)$ which is then amplified by the IF amplifier whose output is fed to the detector. Finally, the audio output voltage is fed to pen recorder wherein the radio signal is received after proper calibration. Figure 2.9 shows a block diagram of the super heterodyne (FM) receiver whose ultimate output is properly tuned at 92.7 MHz. RF voltage amplifier designed for amplification of narrow band of frequencies using tank [46, 47].

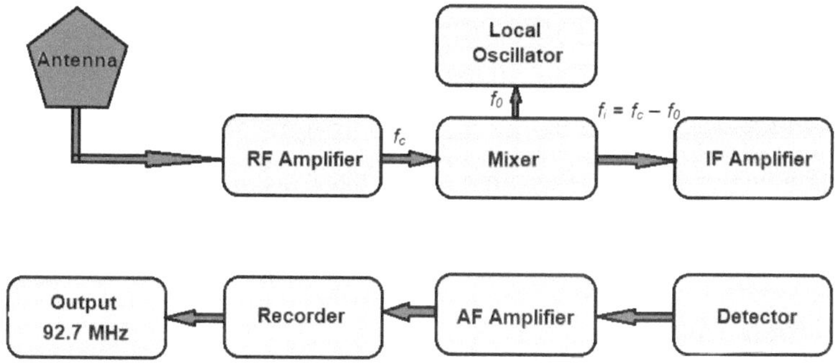

Figure 2.9 Block diagram of the VHF receiver used to receive FM radio signal

2.3 Data Storage Techniques

The received signal through the amplifier is fed either to the digital storage oscilloscope (DSO) connected to the master computer or to the spectrum analyzer for video recording by a digital video camera. When the storage oscilloscope is used for recording, we observe constantly varying signal voltages along the vertical axis ('Y' axis), against the horizontal time axis ('X' axis). Although an oscilloscope displays voltage on its vertical axis, other physical quantity that can be converted to a voltage can be displayed as well [40].

Digital Storage Oscilloscope of GDS-1000 Series we used for recording purposes. In the Digital Storage Oscilloscope (DSO), vertical input is digitized by an analog to digital converter to create a data set which is stored in the memory of a microprocessor. The data set is processed and then sent to the LCD flat panel display. The data set can be sent over a WAN or a LAN for processing or archiving. The screen image can be directly recorded on paper by means of an attached printer or plotter, without the need for an oscilloscope camera. The DSO's own signal analysis software can extract many useful time-domain features (e.g. rise time, pulse width, amplitude), frequency spectra, histograms and statistics, persistence maps, and a large number of parameters meaningful to engineers in specialized fields such as telecommunications, disk drive analysis and power electronics [41].

The received data from LPDA through DSO are subsequently transferred over a LAN for computer simulation. We have also fed the receiving signal to the input of the spectrum analyzer. The data so recorded in the spectrum analyzer are then analyzed either by computer simulation or by video camera [42].

References

1. S. Joardar, S. Bhattacharyya, A. B. Bhattacharya and C. R. Datta, "Radio Astronomy and Supersynthesis: A Survey," *Progress in Electromagnetics Research B*, 22, 73-102, 2010.

2. A. B. Bhattacharya, *Search for Extraterrestrial Intelligence through Radio Astronomical Studies*, Lambert Academic Publishing, Germany, 325-345, 2012.

3. A. B. Bhattacharya, S. S. Banerjee and R. Bhattacharya, "A Roadmap for detecting Extraterrestrial Intelligent Life," *PIERS Proceedings*, Beijing, China, March 23-27, pp. 1400-1403, 2009.

4. A. Alxander, "Space Topics: Extrasolar Planets Astrometry: The Past and Future of Planet Hunting", The Planetary Society, 2006.

5. A. Wolszczan and D. A. Frail, "A Planetary System Around the Millisecond Pulsar PSR1257+12", *Nature,* 355, 145-147, 1992.

6. E. R. Williams, S. G. Geotis, and A. B. Bhattacharya, "A Radar Study of the Plasma and Geometry of Lightning", *Journal of the Atmospheric Sciences*, 46, 1173-1185, 1989.

7. W. D. Heintz, "The binary star 70 Ophiuchi revisited", *Journal of the Royal Astronomical Society of Canada*, 82, 140-145, 1988.

8. C. A. Scharf, *Extrasolar planets and astrobiology*, University science books, Sausalito, California, 2009.

9. E. Agol, J. Steffen, R. Sari and W. Clarkson, "On detecting terrestrial planets with timing of giant planet transits", *Monthly Notices of the Royal Astronomical Society*, 359, 567-579, 2005.

10. J. S. Greaves, M. C. Wyatt, W. S. Holland and W. F. R. Dent, "The Debris Disk around Tau Ceti: A Massive Analogue to the Kuiper Belt", *Monthly Notices of the Royal Astronomical Society*, 351, 54–58, 2004.

11. E. Chaisson, *The Life Era: Cosmic Selection and Conscious Evolution*, New York, Atlantic Monthly Press, 1987.

12. M. Crowe, *The Extraterrestrial Life Debate (1750-1900), The Idea of a Plurality of Worlds from Kant to Lowell*, Cambridge University Press, 1986.

13. J. Schneider, The Extrasolar Planets Encyclopedia, 2006.

14. J. A. Angelo, Jr. The Extraterrestrial Encyclopedia: Our Search for Life in Outer Space, New York: Facts on File, 1985.

15. N. Calder, Timescale: An Atlas of the Fourth Dimension, New York, 1983.

16. K. L. Luhman, L. Adame, P. D'Alessio and N. Calvet, "Discovery of a Planetary-Mass Brown Dwarf with a Circumstellar Disk", *The Astrophysical Journal,* 635, 93–96. 2005.

17. J. M. Jenkins and R. D. Laurance, "Detecting Reflected Light from Close in Giant Planets Using Space-based Photometers", *The Astrophysical Journal,* 595, 429-445, 2003.

18. S. J. Garber, "Searching for Good Science: The Cancellation of NASA's SETI Program", *Journal of the British Interplanetary Society,* 52, 3-12, 1999.

19. H. I. Ewan and E. M. Purcell, "Observation of a line in the galactic radio spectrum Radiation from Galactic Hydrogen at 1,420 Mc./sec". *Nature*, 168, 356, 1951.

20. W. N. Christiansen and J. V. Hindman, "A preliminary survey of 1420 Mc/s line emission from galactic hydrogen", *Australian Journal of Scientific Research*, A5, 437, 1952.

21. C. A. Muller and J. H. Oort, "The Interstellar Hydrogen Line at 1,420 Mc/sec., and an Estimate of Galactic Rotation", *Nature*, 168, 357–358, 1951.

22. S. A. Wouthuysen, "On the excitation mechanism of the 21 cm interstellar hydrogen emission line", *Physica*, 18, 75–76, 1952.

23. S. A. Wouthuysen, "On the excitation mechanism of the 21-cm (radio-frequency) interstellar hydrogen emission line", *Astronomical Journal*, 57, 31-32, 1952.

24. E. Kallas and W. Reich, "A 21 CM radio continuum survey of the galactic plane between L equals 93 deg and L equals 162 deg", *Astronomy and Astrophysics Supplement Series*, 42, 227-243, 1980.

25. J. H. Oort, "Measures of the 21-cm Line Emitted by Interstellar Hydrogen," *Vistas in Astronomy* (*ELSEVIER*), 1, 607-616, 1955.

26. W. Reich, "A radio continuum survey of the northern sky at 1420 MHz. I", *Astronomy and Astrophysics Supplement Series*, 48, 219-297, 1982.

27. Radio Astronomy Supplies – Spectra-Cyber™ - Technical Manual Version 1.6, 2007.

28. R. W. Fairbridge, "The Encyclopedia of Atmospheric Sciences and Astrogeology". *Encyclopedia of Earth Sciences Series*, Reinhold Publishing Corporation, 1214 p, New York, 1967.

29. N. F. Arnold, and T. R. Robinson, "Solar magnetic flux influences on the dynamics of the winter middle atmosphere". *Geophysical Research Letters*, 28, 2381–2384, 2001.

30. D. Grubor, D. Sulic, and V. Zigman, "Influence of Solar X-ray flares on the earth-ionosphere waveguide", *Serbian Astronomical Journal,* 171, 29-35, 2005.

31. A. B. Bhattacharya, S. Joardar, R. Bhattacharya, *Astronomy and Astrophysics*, Infinity Science Press, 2008.

32. A. O. Benz, C. Monstein, H. Meyer, CALLISTO, A New Concept for Solar Radio Spectrometers, Kluwer Academic Publishers, The Netherlands, 2004.

33. K. Rohlfs, T. L. Wilson, *Tools of Radio Astronomy*, Springer, 2006.

34. D. E. Isbel, "Log periodic dipole antenna", *IRE Trans. Antenna & Propagation*, AP-8, 260-267, 1960.

35. G. H. Zhai, W. Hong, K. Wu, Z. Q. Kuai, "Wideband substrate integrated printed log-periodic dipole array antenna", *IET Microwaves, Antennas & Propagation*, 4, 899–905, 2010.

36. A. B. Bhattacharya, S. Joardar, A. Nag, K. Roy, "Design parameters of a log periodic dipole array and associated characteristics", *Proc. National Conference on Materials,*

Devices and Circuits in Communication Technology, IETE Burdwan Sub-Center, 10-11, 2010.

37. K. Mohammadpour-Aghdam, R. Faraji-Dana, J. Rashed-Mohassel, "Compact dual polarisation planar log-periodic antennas with integrated feed circuit", *IEE (IET) Proc. Microwaves, Antennas and Propagation Journal*, 152, 359-366, 2005.

38. A. G. Roederer, "Calculations of the Electromagnetic field radiated by a log periodic dipole antenna", *Philips research reports*, 25, 175 - 188, 1970.

39. P. L. Cupitt, 'Digital Storage Oscilloscope' University Of Durham, 1998.

40. B. Gottfried, 'Schaum's Outlines - Programming With C' McGraw-Hill, 1996.

41. M. Se-yuen, "Wave experiments using low-cost 40 kHz ultrasonic transducers", Physics Education, 38, 441-446, 2003.

42. X. M. Shao and A. Jacobson, "Model Simulation of Very Low-Frequency and Low-Frequency Lightning Signal Propagation Over Intermediate Ranges", IEEE Electromagnetic Compatibility Society, 51, 519 – 525, 2009.

43. R. Dowden, et al, "World-wide lightning location using VLF propagation in the Earth-ionosphere waveguide", Antennas and Propagation Magazine, IEEE, 50, 40 – 60, 2008.

44. C. A. Balanis, *Antenna Theory, Analysis and Design, 3rd Edition*, John Wiley & Sons, Inc., Hoboken, New Jersey, 2005.

45. D. Senić, A. Šarolić, "Simulation of a shipboard VHF antenna radiation pattern using a complete sailboat model", SoftCOM 2009 17. International Conference on Software, Telecommunications & Computer Networks.

Chapter 3

INVESTIGATION OF INTERSTELLAR SIGNAL USING H-LINE SPECTROMETER

3.1 Background

An extrasolar planet or exoplanet is a planet outside the solar system. Most of the exoplanets are massive giant planets similar to that of Jupiter [1-3]. There is a serious limitation in the methods of detecting extrasolar planets. Any planet is an extremely faint light source compared to its parent star. Light from the parent star causes a glare that washes it out. For this limitation only a very few extrasolar planets have been detected directly [4-8]. Generally, the life, which originates outside the earth, signified as extraterrestrial life. Tentative types of extraterrestrial life are in the range between sapient beings to bacterial life. Since convincing evidence of life is not extensively acknowledged by the scientific community, its existence is still hypothetical._ Scientists and general public have realized that probably we are not the only civilization in our galaxy. There may even contain dozens or hundreds of civilizations scattered among its 400,000,000,000 stars. The search for extraterrestrial intelligence (SETI) has now become a mainstream within the scientific enterprise [1, 9]. In fact, efforts of radio astronomy to detect a signal or message from another civilization are increasing rapidly. Some of the events in the entire sweep of human history would be as significant and far-reaching, affecting our deepest beliefs about the nature of the universe, our place and role in it and what lies ahead for human civilization. Reasonably the top priority in the current agenda of our civilization is to seek contact and to prepare for successful interaction with intelligent life from somewhere else in our galaxy. The purpose of this chapter is to present a method of new radio technique and an outline of the instruments required for the purpose that can be implemented for interstellar studies.

3.2 Early Searches

At N.R.A.O., Green Bank, with a radio telescope (diameter of antenna 26 m) Drake [10] took first attempt to search extraterrestrial life. With a single tunable channel of 100 Hz bandwidth at the hydrogen line corresponding to 1.4 GHz, it was a 'Targeted Search' for two stars, 'Tau Ceti' and 'Epsilon Eridani'. At Ohio State University with 110 m radio telescope Dixon [11] continued observation for the same purpose while with 26 m radio telescope an ultra-narrow band multi-channel search was made by Horowitz and Forster [12]. The summary of all these early attempts is presented in Table 3.1.

Table 3.1 Summary of early attempts of interstellar communication

Type of Search	Location	Scientist	Year	Antenna Diameter
Targeted search	N.R.A.O., Green Bank	Frank Drake	1960	26m
All sky search	Ohio State University	Robert Dixon	1985	110m
All sky search	Harvard University	Paul Horowitz & J. Forster	1985	26m

3.3 Confirmed Extrasolar Planets

In 1995, the first exoplanets were detected. It is 51 Pegasi b, it is a hot Jupiter with a 4.2-day orbit revolution around the star 51 Pegasi [13]. A list of confirmed extrasolar planets as discovered during the last twenty-five years is presented in Figure 3.1 in the form of histogram. In 2014, almost greater than 800 exoplanets have been discovered, out of which more than 720 planets have been discovered through transit time method, while Table 3.2 to table 3.4 provide the latest exoplanets discovered in 2015 (up to February). This map (Figure 3.2) shows all of the star systems within 14 light-years of the Sun (shown as *Sol*), except for four brown dwarfs discovered after 2009. Double and triple stars are shown "stacked", but the true location is the star closest to the central plane [14].

Table 3.2 Exoplanets discovered in 2015 (with their mass compared to Jupiter) [13]

Star Name	Planet Name	Mass (Compared to Jupiter)	Radius (Compared to Jupiter)
HD 145934	b	2.28	
EPIC 201367065	b		0.191
EPIC 201367065	c		0.153
EPIC 201367065	d		0.136
KOI-4427	01		0.164
KELT-7	b	1.28	1.533
Kepler-436	b		0.244

Kepler-437	b		0.191
Kepler-438	b		0.0999
Kepler-439	b		0.2
Kepler-440	b		0.166
Kepler-441	b		0.146
Kepler-442	b		0.12
Kepler-443	b		0.208
Kepler-444	b		0.036
Kepler-444	c		0.0443
Kepler-444	d		0.0473
Kepler-444	e		0.0487
Kepler-444	f		0.0661
Kepler-445	b	0.02	0.141
Kepler-445	c	0.027	0.224
Kepler-445	d	0.011	0.112
Kepler-446	b	0.014	0.134
Kepler-446	c	0.009	0.098
Kepler-446	d	0.01	0.12

Table 3.3 Exoplanets discovered in 2015 (with their orbital period, semi major axis and detection type) [13]

Star Name	Planet Name	Orbital Period (Day)	Semi Major Axis	Detection Type
HD 145934	b	2730	4.6	Radial Velocity
EPIC 201367065	b	10.05403	0.0769	Transit
EPIC 201367065	c	24.6454	0.1399	Transit
EPIC 201367065	d	44.5631	0.2076	Transit
KOI-4427	01	147.6606	0.419	Transit
KELT-7	b	2.734775	0.04415	Transit
Kepler-436	b	64.00205	0.339	Transit
Kepler-437	b	66.65062	0.288	Transit
Kepler-438	b	35.23319	0.166	Transit
Kepler-439	b	178.1396	0.563	Transit
Kepler-440	b	101.1114	0.242	Transit
Kepler-441	b	207.2482	0.64	Transit

Kepler-442	b	112.3053	0.409	Transit
Kepler-443	b	177.6693	0.495	Transit
Kepler-444	b	3.600011	0.04178	Transit
Kepler-444	c	4.545884	0.04881	Transit
Kepler-444	d	6.189392	0.06	Transit
Kepler-444	e	7.743493	0.0696	Transit
Kepler-444	f	9.740486	0.0811	Transit
Kepler-445	b	2.984151		Transit
Kepler-445	c	4.871229		Transit
Kepler-445	d	8.15275		Transit
Kepler-446	b	1.565409		Transit
Kepler-446	c	3.036179		Transit
Kepler-446	d	5.148921		Transit

Table 3.4 Exoplanets discovered in 2015 (with their Eccentricity and Discovery year) [13]

Star name	Planet name	Eccentricity	Discovery year
HD 145934	b	0.053	2015

EPIC 201367065	b		2015
EPIC 201367065	c		2015
EPIC 201367065	d		2015
KOI-4427	01	0.02	2015
KELT-7	b		2015
Kepler-436	b	0.19	2015
Kepler-437	b	0.02	2015
Kepler-438	b	0.03	2015
Kepler-439	b	0.03	2015
Kepler-440	b	0.34	2015
Kepler-441	b	0.1	2015
Kepler-442	b	0.04	2015
Kepler-443	b	0.11	2015
Kepler-444	b	0.16	2015
Kepler-444	c	0.31	2015
Kepler-444	d	0.18	2015
Kepler-444	e	0.1	2015
Kepler-444	f	0.29	2015

Kepler-445	b	0	2015
Kepler-445	c	0	2015
Kepler-445	d	0	2015
Kepler-446	b	0	2015
Kepler-446	c	0	2015
Kepler-446	d	0	2015

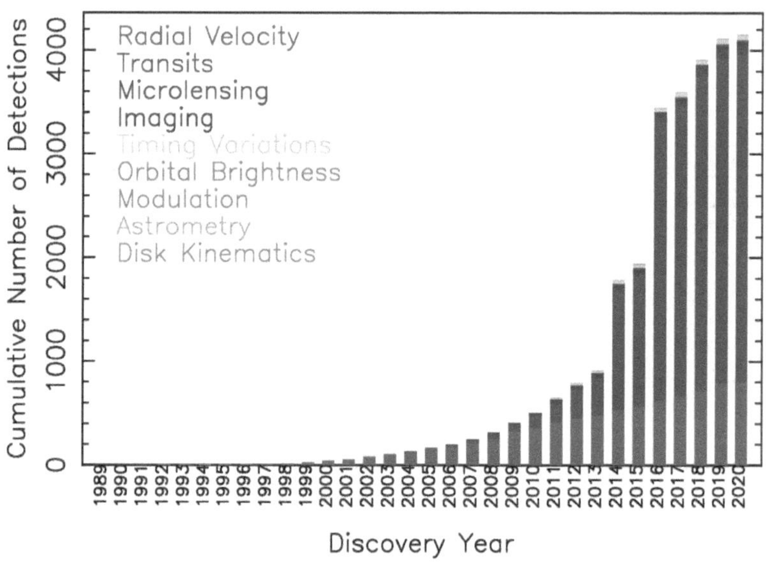

Figure 3.1 Confirmed exoplanets during 1989 to 2020 [13]

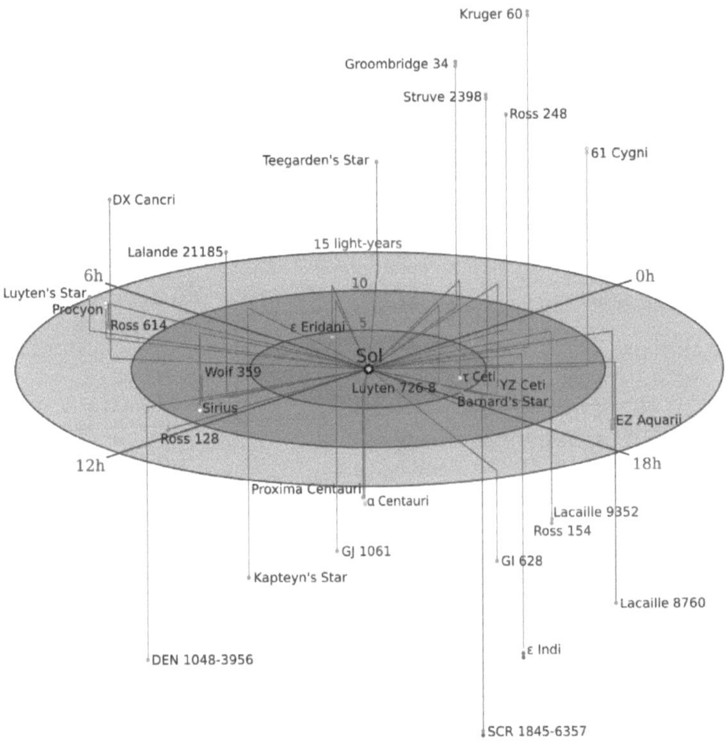

Figure 3.2 Maps of nearby stars [14]

3.4 Present Approach and Theoretical Considerations

It was proposed [15, 16] that sky noise windows may be taken as an appropriate alternative search strategy for interstellar communication. Considering the sky noise windows of the atmosphere we are suggesting the radio communication approach here. A sky noise diagram, indicating sky noise temperature against frequency, is shown in Figure 3.3. If the hydrogen line and the water line are considered, we find that the hydrogen (H) line is at 21 cm whereas the hydroxyl radical (OH) line is at 18 cm. When hydrogen is combined with hydroxyl radical, it produces water (H_2O) whose wavelength lies between 18 cm and 21 cm. The 21cm line is formed by an exciting quantum peculiarity in the nature of the electrons within the hydrogen atom. There also exist 2 mm line of start of water molecule extending up to 14 mm line provides a second water hole [17]. Table 3.5 provides the outline of the interstellar components and their corresponding states with their fractional volume, density and temperature.

Table 3.5 (a) Interstellar Components with their Fractional Volume and Temperature (K)

Component	Fractional Volume	Temperature (K)
Molecular clouds	< 1%	20 – 50
Cold Neutral Medium (CNM)	1 – 5%	50 – 100
Warm Neutral Medium (WNM)	10 – 20%	1000 – 5000
Warm Ionized Medium (WIM)	20 – 50%	$10^3 – 10^4$
H II regions	~ 10%	10^4
Coronal gas		
Hot Ionized Medium (HIM)	30 – 70%	10^6 - 10^7

Table 3.5 (b) Interstellar Components with their Density and State

Component	Density (atoms/cm^2)	State
Molecular clouds	10^3 - 10^6	Hydrogen molecules
Cold Neutral Medium (CNM)	$1 – 10^3$	Neutral hydrogen atoms
Warm Neutral Medium (WNM)	$10^{-1} – 10$	Neutral hydrogen atoms
Warm Ionized Medium (WIM)	10^{-2}	Ionized hydrogen
H II regions	$10^2 – 10^4$	Ionized hydrogen
Coronal gas		Highly ionized
Hot Ionized Medium (HIM)	$10^{-4} – 10^{-2}$	Both hydrogen and trace metal

This window is selected as the first water hole. To make radio link [18] this window can be exploited by galactic civilizations and also our own civilization with each other. Again, there is a water molecule which starts from 2mm extend up to 14 mm line and this may be considered as a second "water hole". Let us consider a radio transmitter installed at earth's surface, isotropically radiates a power P_t over a transmitter bandwidth B_t. Figure 3.4(a) illustrates the principle of the radio communication link, while Figure 3.4(b) reveals a block diagram of the spectrograph including low noise amplifier, transmitter and receiver.

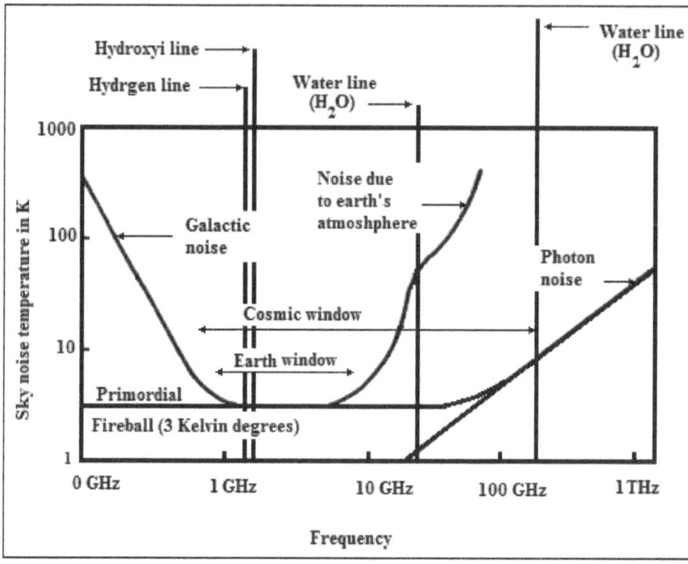

Figure 3.3 Sky noise windows ("water holes"), may be taken as most appropriate for interstellar communications

The flux density generates at a distance *d* is $(P_t/4\pi d^2 B_t)$. Let A_{er} be the effective aperture of the receiving antenna, then power it receives at a distance d is

$$P_r = \frac{P_t A_{er} B_r}{4\pi d^2 B_t} \qquad (3.1)$$

[where, $B_r =$ receiving bandwidth]

If the directivity of the transmitting antenna be *D*, the received power becomes,

$$P_r = D\frac{P_t A_{er} B_r}{4\pi d^2 B_t} \qquad (3.2)$$

Now the value of *D* is,

$$D = \frac{4\pi A_{et}}{\lambda^2} \qquad (3.3)$$

Putting the value of D, we have,

$$P_r = \frac{4\pi A_{et}}{\lambda^2} \cdot \frac{P_t A_{er} B_r}{4\pi d^2 B_t} = \frac{A_{et} P_t A_{er} B_r}{\lambda^2 d^2 B_r} \quad (3.4)$$

[where, A_{et} = effective aperture of the transmitting antenna,

λ = wavelength]

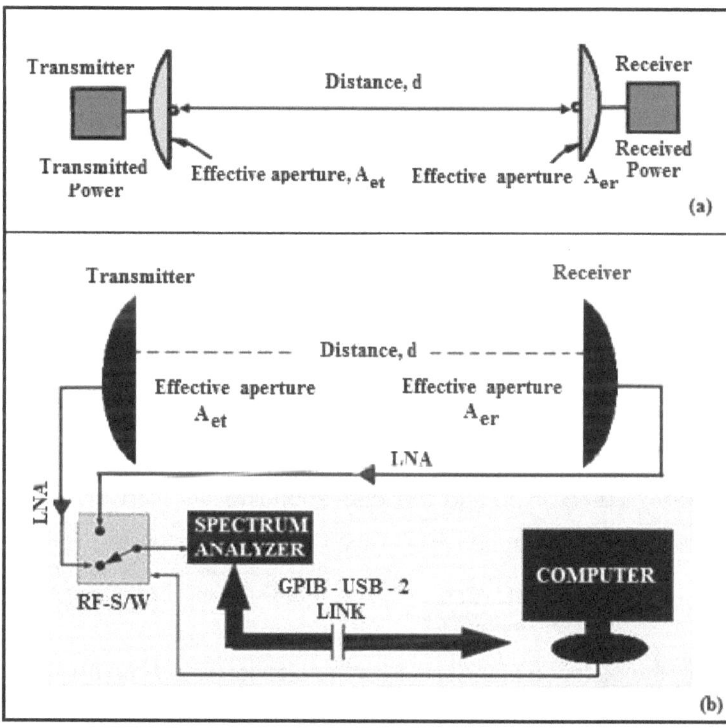

Figure 3.4 (a) Principle of radio transmission and reception for exoplanet link; (b) Suggested technique to be implemented

The received power is suitable for communication or not, determined by measuring signal to noise power ratio *S/N*. Applying Nyquist relation, the noise power can be written as,

$$P_n = kT_{sys}B_r \quad (3.5)$$

[where, k = Boltzmann constant and T_{sys} = system temperature]

Now taken the ratio of equations (3.4) and (3.5) we get

$$\frac{P_r}{P_n} = \frac{A_{er}P_tA_{et}B_r}{kT_{sys}B_r\lambda^2d^2B_t} \quad (3.6)$$

Now according to statistical point of view, the noise can be making smaller by integrating over the bandwidth B_r for a time t, which can be expressed as $\sqrt{B_rt}$.

Therefore, from equation (3.6) we have,

$$\frac{S}{N} = \frac{P_r}{P_n/\sqrt{B_rt}} = \frac{A_{et}P_tA_{er}\sqrt{B_rt}}{kT_{sys}\lambda^2d^2B_t} \quad (3.7)$$

If in such case when the bandwidths of the communicative signal which transmitted by the earth are matched with the extragalactic receiver i.e. $B_r = B_t$ and the corresponding duration of pulse is matched to bandwidth i.e. $B_rt = 1$ then the value of *S/N* is

$$\frac{S}{N} = \frac{A_{et}P_tA_{er}}{kT_{sys}B_r\lambda^2d^2} \quad (3.8)$$

Hence, the distance of the radio communication link *d* can be derived from,

$$d = \sqrt{\frac{P_tA_{et}A_{er}}{kT_{sys}(S/N)\lambda^2B_r}} \quad (3.9)$$

From above equation we see that the range of *d* will be increased with the decreased value of B_r. Therefore, for better communication the small bandwidth should be used. From some significant considerations, Drake and Helou [19] concluded that due to time broadening by multiple scattering of the interstellar medium, a minimum usable bandwidth is 0.1 Hz approximately.

3.5 Interstellar Links to Extraterrestrial Civilization

Let us consider a hypothetical interstellar communication link for the purpose of extraterrestrial observations. We choose two dishes of diameters 75 m and 150 m respectively for using signal transmission and reception independently for two different situations. The effective apertures of the antennas can be calculated from: $A_{et} = A_{er} = 0.125\pi$ (diameter of the dish)^2. If we take, $S/N = 1$, $k = 1.38 \times 10^{-23}$ JK^{-1}, T_{sys} = 10 K and B_r = 0.1 Hz, then from equation (3.9), we can calculate the maximum distance of the transmitted signal in terms of the wavelength (λ). This is shown in the Table 3.6.

Table 3.6 Dish diameters, transmitted powers and corresponding distances

Dish Diameter in (m)	$\boldsymbol{A_{er}=A_{et}}$ **in (m^2)**	**Transmitted power (P_t) in (W)**	***d* from equation (3.9) (×10^{16}) in (m)**
75	2208.9	10^7	$\frac{188.0}{\lambda}$
		10^5	$\frac{18.8}{\lambda}$

150	8835.7	10^7	$\frac{752.1}{\lambda}$
		10^5	$\frac{75.2}{\lambda}$

We choose the two water holes, as indicated in Figure 3.3, for the selection of frequency. The wavelengths (also the corresponding frequencies) and distances of the radio link for the two holes are specified in Table 3.7 to table 3.10.

Table 3.7 First Water hole with 18.0 cm wavelength

Frequency (in GHz)	Dish Diameter in (m)	Transmitted power (P_t) In (w)	Distance(d) of radio link (in m) ($\times 10^{16}$); [using Table 3.4]
1.67	75	10^7	1044.6
		10^5	104.7
	150	10^7	4178.6
		10^5	417.8

Table 3.8 First Water hole with 21.0 cm wavelength

Frequency (in GHz)	Dish Diameter in (m)	Transmitted power (P_t) In (w)	Distance(d) of radio link (in m) ($\times 10^{16}$); [using Table 3.4]
1.43	75	10^7	895.4
		10^5	89.5
	150	10^7	3581.6
		10^5	358.2

Table 3.9 Second Water hole with 2.0 mm wavelength

Frequency (in GHz)	Dish Diameter in (m)	Transmitted power (P_t) In (w)	Distance(d) of radio link (in m) ($\times 10^{16}$); [using Table 3.4]
150.00	75	10^7	94018.5
		10^5	9401.5
	150	10^7	376073.0
		10^5	37607.0

Table 3.10 Second Water hole with 14.0 mm wavelength

Frequency (in GHz)	Dish Diameter in (m)	Transmitted power (P_t) In (w)	Distance(d) of radio link (in m) ($\times 10^{16}$); [using Table 3.4]
21.43	75	10^7	13431.2
		10^5	1343.1
	150	10^7	53724.7
		10^5	5372.4

From this table it is clear that the two water holes have frequencies 1.67 to 1.43 GHz (first water hole) and 21.43 to 150.00 GHz (second water hole) respectively. In order to calculate *d*, we assumed that the dish diameter of the receiving antenna on the earth is identical. And their aperture efficiency is 50% with the system temperature at the earth station. Take this system temperature T_{sys} is 10 K with the corresponding bandwidth 0.1 Hz.

Now 1 Ly being equal to 10^{16} m (approximately), the corresponding distances become 1044.65 Ly, 104.46 Ly, 4178.59 Ly, 417.85 Ly (for wavelength18 cm); 895.41 Ly, 89.53 Ly, 3581.65 Ly, 358.16 Ly (for wavelength 21 cm); 94018.50 Ly, 9401.50 Ly, 376073.00 Ly, 37607.00 Ly (for wavelength 2 mm) and 13431.21 Ly, 1343.07 Ly, 53724.71 Ly, 5372.43 Ly (for wavelength 14 mm). If for simplicity we take the uniform star density as 0.01 per Ly^3 then by measuring the volume for different radii and directivities of corresponding antenna we can calculate the number of stars to be recognized by the proposed antenna beam. The probable number of stars as derived from this suggested technique is presented in Table 3.11.

Here volume $V = \frac{4}{3}\pi\ (Radius\ in\ Ly)^3$

And directivity of the transmitting antenna $D = \frac{4\pi A_{et}}{\lambda^2}$

Therefore, number of stars = Volume/Directivity

$$= \frac{V}{D} = \frac{(Radius\ in\ Ly)^3 \times \lambda^2}{3A_{et}}$$

In the above calculation we find that depending on the transmitted power and hence on the radius (in Ly) the numbers of stars vary. It is interesting to note that when the radius is significantly small, the number of stars may even be less than one (as shown by * marks) in the table 3.11. This indicates that a minimum value of the transmitted power is essential for detecting and identifying any star.

Table 3.11 The probable number of stars to be identified from the suggested technique

Wavelength (λ)	**Radius (in Ly)**	**Number of stars = Volume/Directivity**
18cm	1044.6	55
	104.5	0.05*

	4178.6	891
	417.8	0.89*
21cm	895.4	47
	89.5	0.05*
	3581.6	764
	358.2	0.76*
2mm	94018.5	5016
	9401.5	5
	376073.0	80262
	37607.0	80
14mm	13431.2	716
	1343.1	0.72*
	53724.7	11466
	5372.4	11

3.6 Further Approach and Results

In this analysis a good number of Star/Constellation reported up to the recent years [20-25] with their corresponding distances have been taken into account for calculating the time taken by the radio signal to reach the Star from the Earth as well as the total time taken by the signal to back to the Earth again [26-27]. The number of probable responding stars at different distances by the varying transmitted power has been critically examined by considering the two "water holes" and applying the standard principle of radio signal transmission [28-29].

We have considered quite a good number of reported Star/Constellation [30] and their corresponding distances for the present investigation as a first step. Using the values of the distances of the stars given in light year (Ly), we have calculated the time taken by the radio signal to reach the concerned Star and also the total time taken by the signal to back to the Earth again.

Figure 3.5 reveals the occurrences of selected stars at different distances while Figure 3.6 presents a scatter plot of the different time zones required by the radio signal to arrive at the earth as echo signal. It appears from Figure 3.5 that more than 60% of stars are situated up to a distance of 500 Ly. However, when Figure 36 is taken into analysis no prominent conclusion can be drawn except that the required time of arrival of the echo signal within 10 years period from its time of transmission is not sufficient in number. This prompted us to re-examine the response probability of the stars by transmitting radio signals through the two water holes using different transmitting powers.

The probability of getting response from the number of stars by transmitting radio signals successfully through the two water holes is a function of transmitting power (***P_t***), wavelength (***λ***) of the transmitted signal and distance (***d***) up to which the signal traverses through the windows[4]. Mathematically, for an effective aperture of the transmitting antenna ***A_{et}*** (or the receiving antenna ***A_{er}***) the power received at a distance ***d***, can be obtained from equation 3.4,

$$P_r = \frac{A_{et} P_t A_{er} B_r}{\lambda^2 d^2 B_t}$$

where ***B_r*** or ***B_t*** is the receiving or transmitting bandwidths.

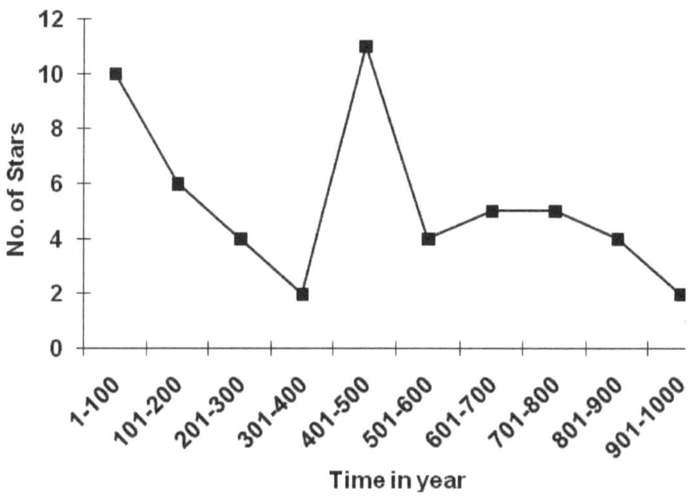

Figure 3.5 Frequency of stars at different distances

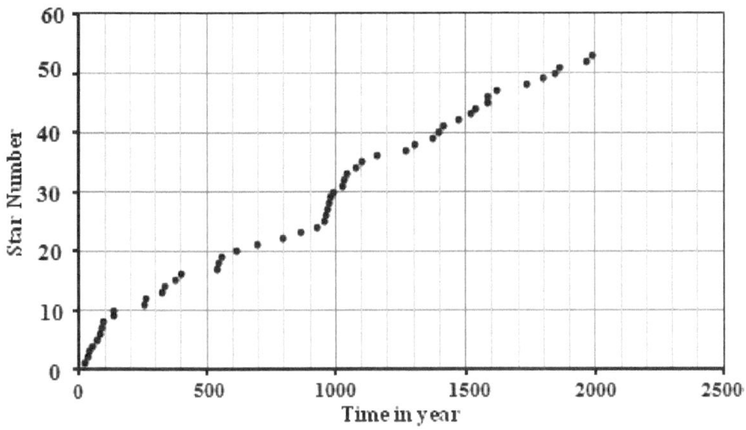

Figure 3.6 Scatter plot, showing the time of arrival of the echo signal

Assuming a minimum bandwidth of the transmitter and receiver same (i.e. $B_r = B_t = 0.1\ Hz$), the probable number of responding stars under varying parameters of P_t, the wave length (λ) and distance (d) are calculated. In course of calculation, we have varied the transmitted power from 10^6 to 10^{18} watt to examine the variation of number of stars for different values of the wavelengths corresponding to the minimum and maximum values of wave lengths for the two water holes.

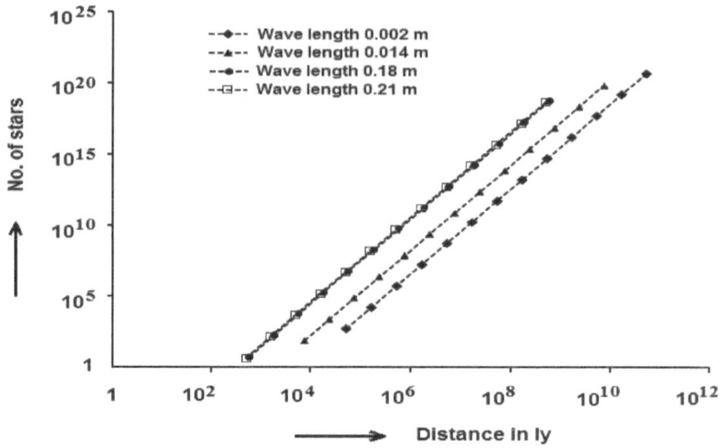

Figure 3.7 Probable numbers of stars at different distances to be detected by radio signal

In Figure 3.7 we have plotted the probable number of stars to be detected by radio signal when transmitted power of the signal through the two 'water holes' are varied over a wide range. It appears from the figure that the number of stars corresponding to the two wavelengths 0.18 m and 0.21 m are almost overlapped and so for the selection of transmitting a signal corresponding to this water hole a mean of these two wavelengths may be chosen as a compromise.

But when the other two wavelengths are taken into consideration it reveals that the probable number of stars become greater at a wavelength of 0.002 m corresponding to higher distances. If, on the other hand, powers in watt are plotted against distances of the probable stars for the four wavelengths corresponding to the two "water holes', we find a linear relationship between the two, as shown in Figure 3.8.

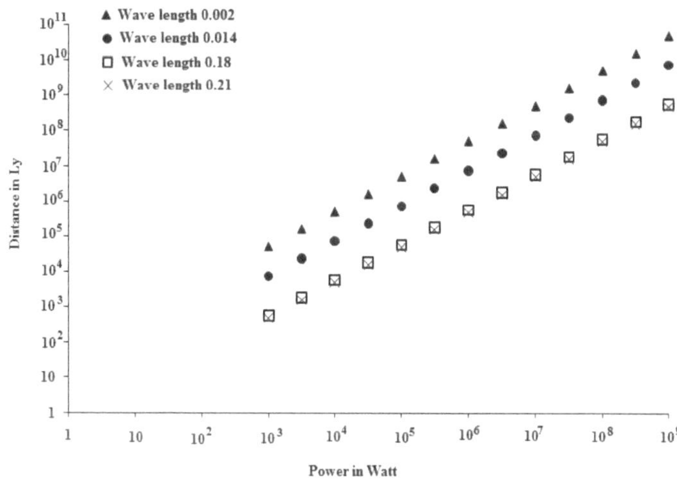

Figure 3.8 Transmitted power plotted against distances of the probable stars corresponding to selected wavelengths

3.7 Discussions

Within this so-called Microwave Window, photons (the substance of electromagnetic communication) travel relatively unimpeded through the interstellar medium, at the speed of light. This is, in fact, the fastest possible speed making photons the fastest spaceships known to man. Thus, the Microwave Window, where natural noise is at a minimum, is a favored region for conducting radio astronomy research, including SETI [31, 32]. Low frequencies are obstructed by galactic noise, primarily due to synchrotron radiation while high frequencies are subject to quantum-effect emissions and the whole continuum experiences a 3 Kelvin background radiation level from the residual radiation of the Big Bang. These natural radiation sources limit our ability for detecting artificial emissions. Moreover, the Earth's own ocean of air generates spectral absorption and emission lines to draw a further curtain across our sky. However, there are a few relatively clear windows on the cosmos. Our eyes evolved to operate in two such windows. These windows will allow us to observe the stars and planets hopefully, if the proposed experiment can be performed with utmost care [1, 9, 23].

The interstellar medium (or ISM) is basically the dust and gas that pervade interstellar space representing the matter that exists between the stars within a galaxy. The energy, in the form of electromagnetic radiation that occupies the same volume is known as the interstellar radiation field. The interstellar medium consists of a dilute mixture of ions, atoms, molecules, larger dust grains, cosmic rays, and also the galactic magnetic fields. It has been well reported that the matter consists of about 99% gas and 1% dust by mass. As a result of primordial nucleosynthesis, the gas is approximately 90% hydrogen and 10% helium by number of nuclei, with an additional heavier elements present in trace amounts, the stars of our Galaxy and presumably the stars in all other billions of galaxies in the Universe are born in interstellar clouds. It is mainly because of that search-strategy planning becomes a very complex one due to so much uncertainty arising for handling various parameters at the time of radio signal transmission through the interstellar medium. This may be minimized by repeating the experiment and giving due importance to the individual parameters taken into account [33, 34].

References

1. D. Charbonneau, L. Allen, S. Megeath, G. Torres, R. Alonso, T. Brown, R. Gilliland, D. Latham, G. Mandushev, F. O'Donovan and A. Sozzetti, "Detection of Thermal Emission from an Extrasolar Planet", *The Astrophysical Journal*, 626, 523 – 529, 2005.

2. D. Deming, S. Seager, J. Richardson and J. Harrington, "Infrared radiation from an extrasolar planet", *Nature*, 434, 740 – 743, 2005.

3. M. Konacki, "Astronomers Confirm the First Image of a Planet Outside of Our Solar System", *ESO*, 2005.

4. E. Chaisson, *The life era: Cosmic selection and conscious evolution*, New York (Atlantic Monthly Press), 1987.

5. M. Crowe, *The extraterrestrial life debate, 1750-1900: The idea of a plurality of worlds from Kant to Lowell*, Cambridge (Cambridge University Press), 1986.

6. B. Campbell, G. A. H. Walker, S. Yang, "A search for substellar companions to solar-type stars", *The Astrophysical Journal*, 331, 902-921, 1988.

7. R. P. Butler, J. T. Wright, G. W. Marcy, D. A. Fischer, S. S. Vogt, C. G. Tinney, H. R. A. Jones, B. D. Carter, J. A. Johnson, C. McCarthy and A. J. Penny, "Catalog of Nearby Exoplanets", *The Astrophysical Journal*, 646, 505–522, 2006.

8. G. Takeda, E. B. Ford, A. Sills, F. A. Rasio, D. A. Fischer and J. A. Valenti, "Structure and Evolution of Nearby Stars with Planets II. Physical Properties of ~ 1000 Cool Stars from the SPOCS Catalog (English)", *California & Carnegie Planet Search*, 2006.

9. D. Charbonneau, T. Brown, A. Burrows and G. Laughlin, *When Extrasolar Planets Transit Their Parent Stars, Protostars and Planets V*, University of Arizona Press, 2006.

10. F. D. Drake, "A reminiscence of Ozma", *Cosmic Search*, 1, 10-15, 1979.

11. R. S. Dixon, "A search strategy for finding extraterrestrial radio beacons", *Icarus*, 20 187-199, 1973.

12. P. Horowitz, and J. Forster, "Project Sentinel: Ultra-Narrowband SETI at Harvard/Smithsonian", *Proc. International Astronomical Union*, 112, 291-303, 1985.

13. http://www.exoplanet.eu/catalog

14. A. B. Bhattacharya, S. S. Banerjee and R. Bhattacharya, "A Roadmap for detecting Extraterrestrial Intelligent Life", *PIERS Proceedings*, Beijing, China, March 23-27, 1400-1403, 2009.

15. A. B. Bhattacharya, S. Joardar and R. Bhattacharya, *Astronomy and Astrophysics*, Jones and Bartlett Publisher, Canada, 2010.

16. J. D. Kraus, *Radio Astronomy*, Cygnus-Quasar books, USA, 12-0 to 12-6, 1986.

17. B. F. Burke and F. G. Smith, *An Introduction to Radio Astronomy, 3rd Edition*, Cambridge University Press, UK, 2010.

18. F. D. Drake and G. Helou, "The optimum frequencies for interstellar communication as influenced by minimum bandwidths", N.A.I.C. (Cornell), Rept.76, 1977.

19. S. Sri Kantha, "Pros and cons in the search for extraterrestrial intelligence", *Medical Hypotheses*, 46, 183–187, 1996.

20. M. J. Fogg, "Temporal aspects of the interaction among the first galactic civilizations: The "interdict hypothesis", *Icarus*, 69, 370–384, 1987.

21. J. Elliott, E. Atwell and B. Whyte, "First stage identification of syntactic elements: An extraterrestrial signal", Proceedings of IAC '2001, The 52nd International Astronautical Congress, October 1-5, IAA-01- IAA.9.2.07, 2001.

22. J. M. Jenkins and R. D. Laurance, "Detecting reflected light from close in giant planets using space-based photometers", *The Astrophysical Journal*, 595, 429-445, 2003.

23. D. Eichler and G. Beskin, "Optical search for extraterrestrial intelligence", *Astrobiology*, 1, 489-493, 2004.

24. J. S. Greaves, M. C. Wyatt, W. S. Holland and W. F. R Dent, "The debris disk around tau Ceti: a massive analogue to the Kuiper Belt", *Monthly Notices of the Royal Astronomical Society,* 351, 54-58, 2004.

25. K. L. Luhman, L. Adame, P. D'Alessio and N. Calvet, "Discovery of a Planetary-Mass Brown Dwarf with a Circumstellar Disk", The Astrophysical Journal, 635, 93-96, 2005.

26. J. Elliott, "A Post-Detection Decipherment Matrix: Acta Astronautica", *Journal of the International Academy of Astronautics*, AA, 2853, 2007.

27. A. B. Bhattacharya, *Search for Extraterrestrial Intelligence through Radio Astronomical Studies*, Lambert Academic Publishing, Germany, 325-345, 2012.

28. J. A. Angelo, Jr., The extraterrestrial encyclopedia: Our search for life in outer space, New York: Facts on File, 1985.

29. A. B. Bhattacharya, S. Sarkar and R. Bhattacharya, "An alternative search strategy for interstellar communication", *Indian Journal of Physics*, 84, 511-515, 2010.

30. E. Herbst, "Chemistry in the Interstellar Medium", *Annual Review of Physical Chemistry*, 46, 27-54, 1995.

31. A. E. Reines and G. W. Marcy, "Optical Search for Extraterrestrial Intelligence: A Spectroscopic Search for Laser Emission from nearby Stars", *Publications of the Astronomical Society of the Pacific*, 114, 416-426, 2002.

32. A. B. Bhattacharya, S. Das, A. Sarkar, M. Debnath and A. Bhoumick, "A new Technique for Investigating Extrasolar Planets", *International Journal of Physics*, 4, 31- 42, 2011.

Chapter 4

OBSERVATION OF GALACTIC RADIO SIGNAL USING H-LINE AT 1420 MHZ

4.1 Background

Radio communication has been attempted by considering the sky noise windows of the atmosphere [1-3] now days. If the hydrogen line and the water line are taken into account one can find that the hydrogen (H) line is at 21 cm whereas the hydroxyl radical (OH) line is at 18 cm. When hydrogen is combined with hydroxyl radical, it produces water (H_2O) whose wavelength lies between 18 cm and 21 cm. The window between 18 cm and 21 cm is treated as the first 'water hole' which when converted to frequency lies between 1420 MHz to 1666 MHz. The chapter reports some interesting features of radio signals at 1420 MHz received from the continuum observations at a tropical station Kalyani (22°59.4'N, 88°26.9'E) [4-6].

The Dutch astronomer Jan Oort learned of Grote Reber's discoveries in radio astronomy and realized that a radio spectral line would be an important tool for discovering the structure of our galaxy. Oort had spent many years studying the rotation and structure of the galaxy using optical means but he was frustrated by the extensive clouds of dust lying in the galactic plane, which block visible light [7]. One can see only a few thousand light years towards the galactic center because the light of distant stars is absorbed. But radio waves can penetrate the dust to show the galactic center and indeed the opposite side of the galaxy. The importance of a spectral line is that the frequency of the line will be shifted due to Doppler Effect which means that the velocity of the gas can be measured. One can then study the differential rotation of the galaxy and estimate distances to gas clouds and thus map the distribution of matter in the galaxy [8].

Oort assigned his student Van de Hulst, with the job of figuring out what radio spectral lines might exist and what their frequencies would be [9]. Since hydrogen is the most abundant element in the universe, he started his studies with hydrogen. He found that a "hyperfine" transition in the ground state of neutral hydrogen would produce radiation in the radio range, at a frequency of 1420 MHz i.e. at a wavelength of about 21 cm. In the ground state of hydrogen, the electron can have its magnetic moment either parallel to that of the proton or anti-parallel. The parallel state has a little more energy and so a transition to the anti-parallel state results in emission of 21 cm radiation.

4.2 Early Attempts to Develop Receiving Equipment

Building a receiver for detecting the predicted line proved to be a daunting task. Grote Reber attempted to design such a receiver in 1947 but set it aside because of changing jobs and moving to the east coast. Dutch engineers worked on the problem but the first successful detection was by Ewen and Purcell at Harvard. In 1950 Ewen and Purcell attempted to build a receiver for detecting the 21 cm line of neutral hydrogen [9]. The original paper by Van de Hulst predicting the existence of the 21 cm line expressed doubt that the line would be detectable. However, a paper by Shklovski in 1948 was more optimistic [9, 10]. Ewen and Purcell assumed that the Dutch were probably not working on detecting the line and that the Russians might soon be. They could probably make the first sensitive detection attempt. It seemed likely that the experiment would yield a negative result but the experience would be worth the trouble and a well-defined upper limit to the 21 cm line emission strength would be important. Ewen proceeded to design the horn antenna, the mixer and the receiver consulting with experts in these fields, including Sam Silver on antenna design and Bob Pound on mixers. The receiver used a frequency switching technique with a view to cancel out background noise, a novel technique for astronomy at the time. The Dutch group implemented Ewen's frequency switching technique and succeeded in detecting the line. A further confirmation of the detection was made by Australian radio astronomers [11, 12]. In the following years, observations of the neutral hydrogen line were made with increasing accuracy by many astronomers, primarily Dutch, American, and Australian. The technique proved to be successful in mapping the large-scale distribution of matter in the galaxy.

4.3 Data Recording Techniques

By clicking we can set the background of the main window which drops a .bmp file for including in publication documents or an animated gif of the data. An X-axis feature permits km/sec for continuum record. Another feature permits to use either dots or connected lines for the plotting. Here X- axis represent frequency scale. The final feature is the Plot Capturing to allow whether or not a bmp graph file is created on each save of the scan. Data stored on disk can be played back in a "movie". Each screen of data is pulled from the disk and rapidly drawn on the Screen. The previous screen is then erased and the next file in sequence is displayed. The files to play are previously stored on the hard disk. The software has the ability to start data collection at a point and this can be chosen at convenient time.

Table 4.1 Specifications of Spectra-Cyber Software

Mode	**Continuum Observation**	**Spectrum Observation**
Auto save	On	On
Noise signal	Off	Off
RE Scan	On	On
Scan Rate	1X	1X
Que	Off	Off
Spec Gain	01	01
Cont. Gain	01	01
Rest Correction	00	00
IF BW (kHz)	15	15
Upper Frequency (kHz)	+0360	+2000
Lower Frequency (kHz)	+0000	-2000
Spec. Int	000.50	000.50
Cont. Int	000.30	000.30

Spec. Offset Voltage (V)	001.00	001.00
Cont. Offset Voltage (V)	000.00	000.00
IF Gain (dB)	017.00	25.00
Time/Step	001.00	001.00

When completed with filling out the settings values the spectrometer moves to Online if the communications channel is open. The final option is the Input Frequency Shift which brings up a dialog box to know how much we want to shift the X- axis values. In Table 4.1 we have presented summary information of the Spectra-Cyber Software.

4.4 Analyses and Results

4.4.1 Data Received in Continuum Mode

We are receiving data of continuum type round-the-clock and noted many interesting characteristics. In analyzing the received continuum data, we have classified them into five categories, *viz.* A, B, C, D and E, to some extent arbitrarily, depending upon their various physical shapes obtained. A typical continuum record at 1420 MHz observed in our laboratory at 17:03:52 UT on May 23, 2013 is exhibited in Figure 4.1. We extract data and then changes the time from UT to IST by adding 5:30 hrs in time scale. We again plot this using Excel file, which is shown in Figure 4.2. Here we noticeably say that the maximum pick arises about 22:35:40 IST. The variations of amplitude occurred from 1.56 volts to almost 1.585 volts.

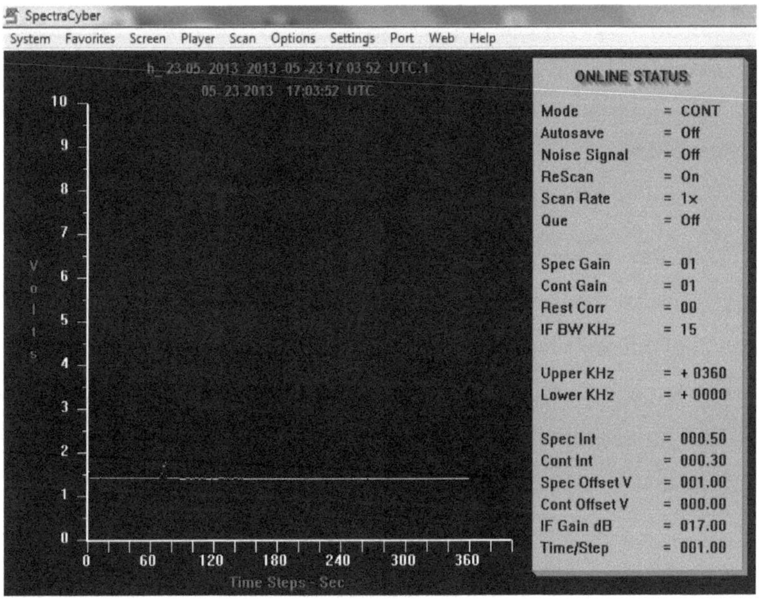

Figure 4.1 A typical continuum records at 1420 MHz recorded in the Laboratory at Kalyani on May 23, 2013. The record reveals a sharp, distinct and prominent pattern appeared in between 17:03:52 UT to 17:10:00 UT.

Some other typical continuum records at 1420 MHz recorded in the Laboratory at Kalyani is illustrated in Figure 4.3 and in Figure 4.4 In both the cases we converted the time scale from UTC to IST. The corresponding dates had indicated in the figures. The physical shapes of the five categories of continuum data following our classification are illustrated in Table 4.2 which clearly shows their shapes and dominant features. We have analyzed 4,59,000 numbers of data and prepared a pie chart showing percentage wise variation of the classified various patterns (Figure 4.5).

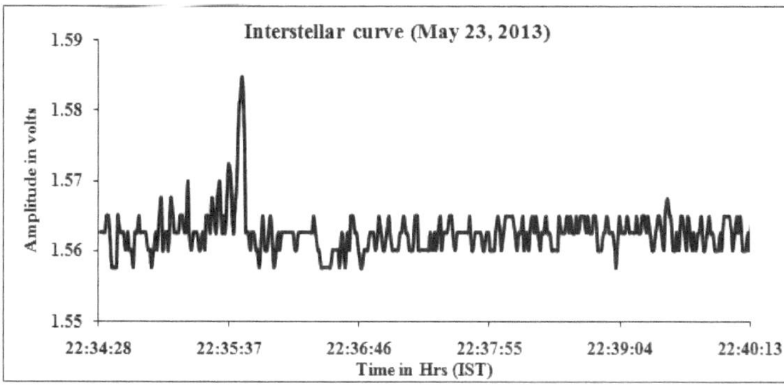

Figure 4.2 Plot of the Interstellar Curve

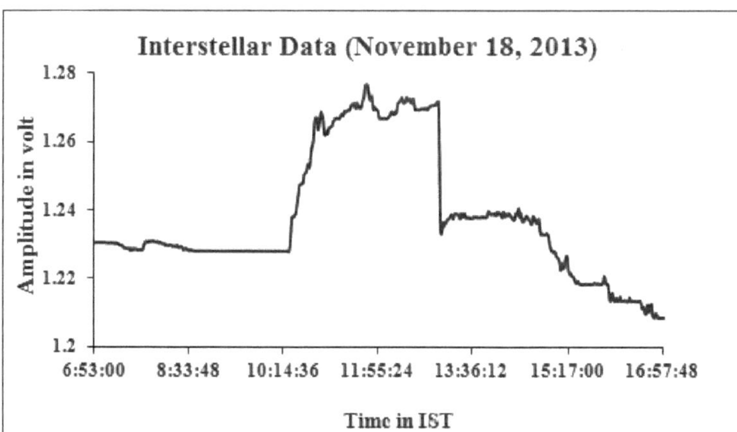

Figure 4.3 Typical variation of interstellar radio signal received at 1420 MHz, on November 18, 2013. (Operating mode is continuum); the horizontal axis representing time scale

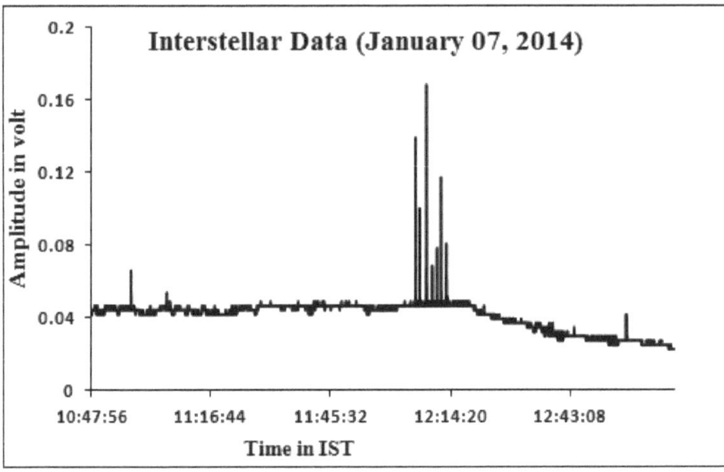

Figure 4.4 Typical variation of interstellar radio signal received at 1420 MHz, on January 07, 2014. (Operating mode is continuum); the horizontal axis representing time scale

Table 4.2 The physical shape of the five categories of continuum data with dominant features

Type	Shape	Dominant features
A	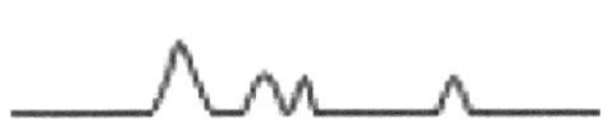	No measurable variation
B		Long time (~20 s) blunt upward peaks

C		Small time (≤10 s), sharp, distinct and prominent upward peaks
D		Long time (~20 s), not so sharp and 'm' shaped upward peaks
E		Sharp downward peaks

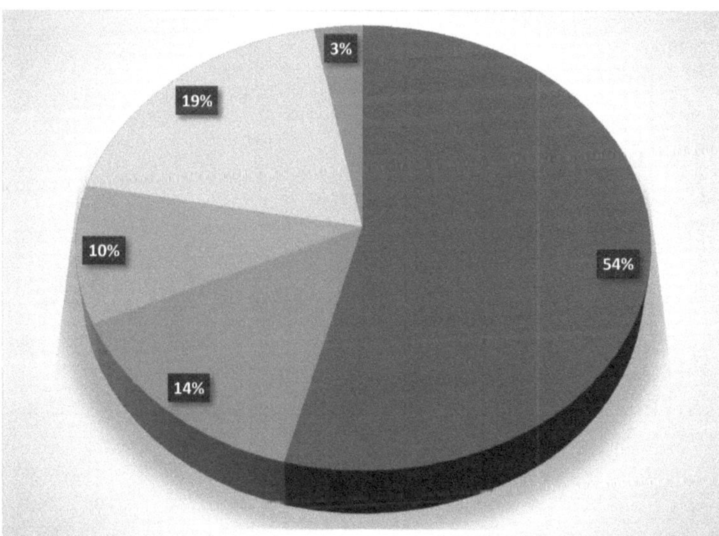

Figure 4.5 A comparative studies of the classified various patterns observed at 1420 MHz

4.4.2 Data Received in Spectrum Mode

When we are receiving data of spectrum type round-the-clock and also noted many interesting characteristics here also. In Figure 4.6 (a) we plot a typical variation of the signal received by 1420 MHz hydrogen line spectra-Cyber receiver (it is a scatter plot that received by the receiver). Here operating mode of the receiver in spectrum and the horizontal axis representing the frequency from 1419 MHz to 1421 MHz. It is interesting to see from the figure that the maximum voltage received at 1420.2 MHz. We extract the data from the save file and again plot this by connecting the dots points, we get another file which shown in the Figure 4.5 (b).

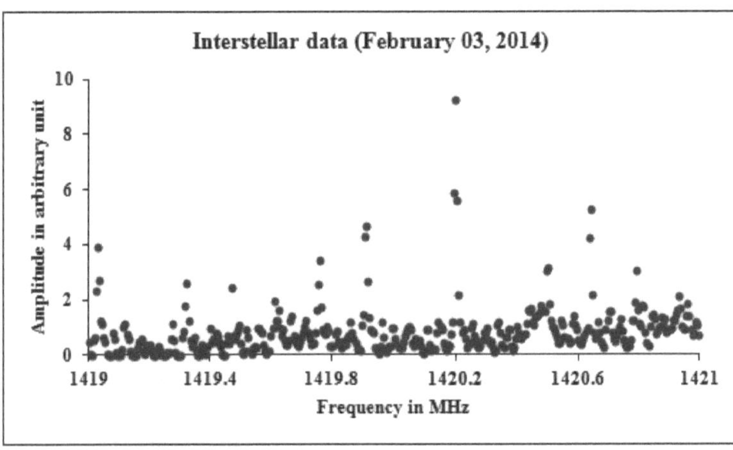

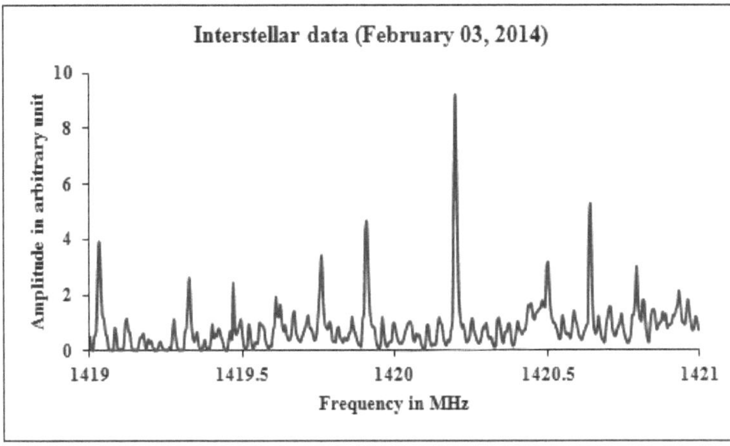

Figure 4.6 Typical variation of interstellar radio signal received at 1420 MHz (operating mode is spectrum); the horizontal axis representing the frequency

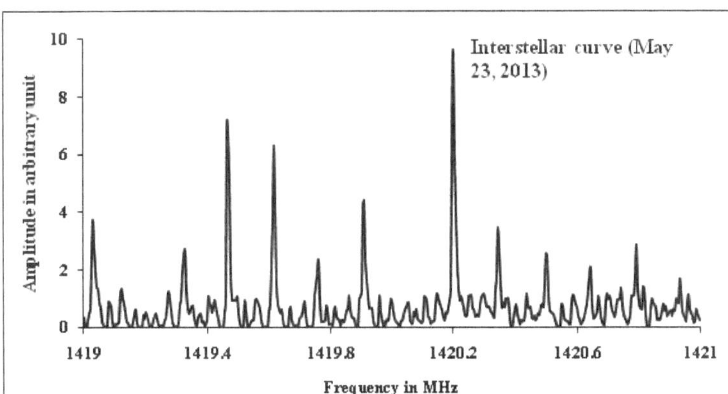

Figure 4.7 Typical variation of interstellar radio signal received at 1420 MHz (operating mode is spectrum); the horizontal axis representing the frequency

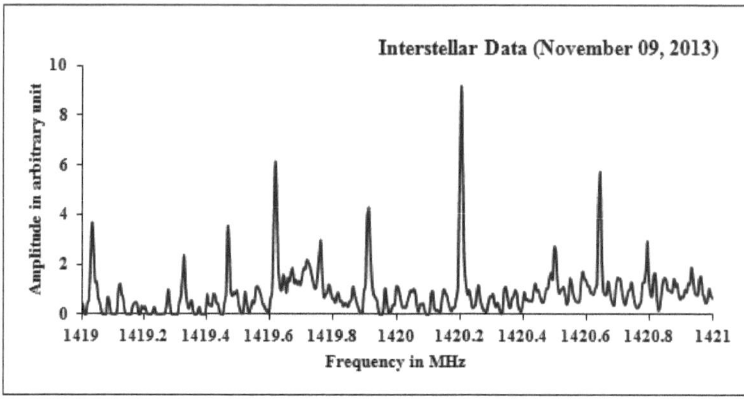

Figure 4.7 Typical variation of interstellar radio signal received at 1420 MHz (operating mode is spectrum); the horizontal axis representing the frequency

Some other data sheet is also present here by Figure 4.7 and Figure 4.8. The corresponding dates the data also indicated in the figurers. It very interesting to note that in all type of data sheet, the maximum pick voltages happened on 1420.2 MHz frequency.

4.5 Conclusions

Several proceed Have been Made in last fifty years to under Standing of the interstellar medium by studying the distribution of 21-cm emission. Absorption By interstellar clouds of radiation emitted by radio sources has also been observed [13-16]. Numerous new radio sources are identified and the catalogue provides a comprehensive summary of both newly discovered and previously known H ii regions and supernova remnants in the outer Galaxy [17-20]. An Interstellar Search System (ISS) is exceptionally vulnerable to interference by manmade coherent radiations and as a matter of fact the terrestrial radio spectrum is largely crowded. For this reason, an extensive strategy is very much essential to hold interference up to a reasonable level. The terrestrial origin of an interfering signal can perhaps be recognized to a high level of certainty by making use of a properly operated ISS.

A simultaneous study is essential to find the degree of using the ISS frequencies particularly where the signals are kept close to the surface of the Earth. In fact, any interstellar searching system is compatible with a large use of the same frequency band when its use is confined to the surface of the Earth. It may be pointed out that the present use of 1400 MHz to 1727 MHz band is more or less compatible with the operation of an ISS, in contrast to bands below 1400 MHz and above 1727 MHz, where powerful aircraft acquisition radars are in operation. There are physical and philosophical arguments which imply that the probability of finding signals from an extraterrestrial civilization is maximized in the frequency band between 1400 MHz and 1727 MHz [1, 21-23]. The 1400 to 1427 MHz band is of interest because interstellar transmissions may take place around the hydrogen line, while the 1427 to 1727 MHz band is located between the hydrogen and hydroxyl lines and lies near the minimum of the noise background. It is to be mentioned here that the 1400 MHz to 1427 MHz is currently allocated exclusively to the radio astronomy service while 1427 to 1727 MHz may be shared with other likely services where their use will not cause any recognized harmful interference during the operation of an ISS [24-25].

References

1. A. B. Bhattacharya, *Search for Extraterrestrial Intelligence through Radio Astronomical Studies*, Lambert Academic Publishing, Germany, 325-345, 2012.

2. M. Konacki, Astronomers Confirm the First Image of a Planet Outside of Our Solar System, ESO, 2005.

3. A. B. Bhattacharya, S. S. Banerjee and R. Bhattacharya, "A Roadmap for detecting Extraterrestrial Intelligent Life", *PIERS Proceedings*, Beijing, China, March 23-27, 1400-1403, 2009.

4. S. A. Wouthuysen, "On the excitation mechanism of the 21 cm interstellar hydrogen emission line", *Physica*, 18, 75–76, 1952.

5. S. A. Wouthuysen, "On the excitation mechanism of the 21-cm (radio-frequency) interstellar hydrogen emission line", *Astronomical Journal*, 57, 31-32, 1952.

6. E. Kallas and W. Reich, "A 21 CM radio continuum survey of the galactic plane between L equals 93 deg and L equals 162 deg", *Astronomy and Astrophysics Supplement Series*, 42, 227-243, 1980.

7. J. H. Oort, "Radio-frequency Studies of Galactic Structure," *Handbuch der Physik*, 53, 100-28, 1959.

8. Oort, “Measures of the 21-cm Line Emitted by Interstellar Hydrogen,” *Vistas in Astronomy* (*ELSEVIER*), 1, 607-616, 1955.

9. Van de Hulst and C. Hendrik, “Radio Waves from Space: Origin of Radiowaves”, Ned. tijd. Natuurkunde, 11, 210-221, 1945. [A portion of this paper is translated and reprinted, with commentary, in Lang, R. Kenneth and O. Gingerich, “A Source Book in Astronomy & Astrophysics”, 1900-1975, Harvard Univ. Press, Cambridge, MA, 1979, 627-32.

10. H. I. Ewan and E. M. Purcell "Observation of a line in the galactic radio spectrum Radiation from Galactic Hydrogen at 1,420 Mc./sec", *Nature*, 168, 356, 1951.

11. W. N. Christiansen and J. V. Hindman, “A preliminary survey of 1420 Mc/s line emission from galactic hydrogen”, *Australian Journal Science* Research, A5, 437, 1952.

12. C. A. Muller and J. H. OOrt, “The Interstellar Hydrogen Line at 1,420 Mc/sec., and an Estimate of Galactic Rotation”, *Nature*, 168, 357–358, 1951.

13. A. R. Taylor, S. J. Gibson, M. Peracaula, P. G. Martin, T. L. Landecker, C. M. Brunt, P. E. Dewdney, S. M. Dougherty, A. D. Gray, L. A. Higgs, C. R. Kerton, L. B. G. Knee, R. Kothes, C. R. Purton, B. Uyaniker, B. J. Wallace, A. G. Willis and D.

Durand, "The Canadian Galactic Plane Survey", *The Astronomical Journal*, 125, 2003

14. R. Paladini, C. Burigana, R. D. Davies, D. Maino, M. Bersanelli, B. Cappellini, P. Platania and G. Smoot, "A radio catalog of Galactic HII regions for applications from decimeter to millimeter wavelengths" *Astronomy and Astrophysics Supplement series*, 397, 213-226, 2003.

15. P. Reich, W. Reich, and E. Fürst, "The Effelsberg 21 cm radio continuum survey of the Galactic plane between l equals 95°. 5 and l equals 240 deg", *Astronomy and Astrophysics Supplement series*, 126, 413-435, 1997.

16. K. Ferriere, "The Interstellar Environment of our Galaxy", *Reviews of Modern Physics*, 73, 1031–1066, 2001.

17. W. Reich, "A radio continuum survey of the northern sky at 1420 MHz. I", *Astronomy and Astrophysics Supplement Series*, 48, 219-297, 1982.

18. T. L. Landecker, P. E. Dewdney, T. A. Burgess, A. D. Gray, L. A. Higgs, A. P. Hoffmann, G. J. Hovey, D. R. Karpa, J. D. Lacey, N. Prowse, C. R. Purton, R. S. Roger, A. G. Willis, W. Wyslouzil, D. Routledge, J. F. Vaneldik, "The synthesis telescope at the Dominion Radio Astrophysical Observatory", *Astronomy and Astrophysics Supplement series*, 145, 509-524, 2000.

19. R. Kothes, K. Fedotov, T. J. Foster and B. Uyaniker, "A catalogue of Galactic supernova remnants from the Canadian Galactic plane survey", *Astronomy and Astrophysics*, 457, 1081-1093, 2006.

20. F. P. Israel, and M. Felli, "Aperture synthesis observations of galactic H II regions. VIII - S106 and S235: Regions of star formation", *Astronomy and Astrophysics*, 63, 325-334, 1978.

21. A. R. Taylor, S. J. Gibson, M. Peracaula, P. G. Martin, T. L. Landecker, C. M. Brunt, P. E. Dewdney, S. M. Dougherty, A. D. Gray, L. A. Higgs, C. R. Kerton, L. B. G. Knee, R. Kothes, C. R. Purton, B. Uyaniker, B. J. Wallace, A. G. Willis and D. Durand, "The Canadian Galactic Plane Survey", *The Astronomical Journal*, 125, 3145–3164, 2003.

22. R. Paladini, C. Burigana, R. D. Davies, D. Maino, M. Bersanelli, B. Cappellini, P. Platania and G. Smoot, "A radio catalog of Galactic HII regions for applications from decimeter to millimeter wavelengths" *Astronomy and Astrophysics Supplement series*, 397, 213-226, 2003.

23. P. Reich, W. Reich, and E. Fürst, "The Effelsberg 21 cm radio continuum survey of the Galactic plane between l equals 95°. 5 and l equals 240 deg", *Astronomy and Astrophysics Supplement series*, 126, 413-435, 1997.

24. V. Bumba, J. Kleczek, J. Olmr and B. Rů ŽiČková-Topolová, "Relation of 1420 MHz radio emission to the distribution of photospheric magnetic fields, coronal emission and filaments", *Bulletin of the Astronomical Institute of Czechoslovakia*, 20, 67-69, 1969.

25. C. R. Kerton, J. Murphy and J. Patterson, "A sharper view of the outer Galaxy at 1420 and 408 MHz from the Canadian Galactic Plane Survey II: the catalogue of extended radio sources", *Monthly Notices of the Royal Astronomical Society,* 379, 289-296, 2007.

Chapter 5

Studying Transient Features Through the Solar Corona by 406.7 MHz Radio Telescope

5.1 Introduction

A coronal mass ejection is a large eruption of coronal plasma and magnetic field rising above the solar corona. The corona is only about one-millionth of the brightness of the Sun. As a matter of fact, its observation, outside eclipses, is quite difficult. Total solar eclipses are rare and also visible from very restricted areas of the Earth. The dynamic pattern of highly structured solar corona is produced owing to dense tube of solar plasma. The solar flare is one of the active phenomena in solar system wherein the magnetic energy is transformed into kinetic energy of the accelerated charged particle through magnetic reconnection [1-3]. The observation of white light corona is a complicated task since many kinds of aberration, largely chromatic aberration may occur and so we need an aberration free instrument in which both spherical as well as chromatic aberration will not be present. But in practice, it is difficult to build a completely aberration free ideal instrument though spherical aberration may be diminished widely. From this consideration, if one use monochromatic beam (i.e. of particular wavelength) then this problem can be minimized. That is, in fact, the main reason for using monochromatic beam in our study. In addition, if we collect monochromatic narrowband radio emission of the corona by using monochromatic filters which allow the particular frequency of the spectrum, which we have chosen for recording, it becomes much simpler [4-6]. At the beginning of 1960's a small sized coronagraph suitable for amateurs was developed. Twenty-five years later the study of the corona has taken on new life with the development radio methods [5, 6]. We have received radio signals from solar corona by using 406.7 MHz radio telescope in our observatory at Kalyani (22.98°N, 88.46°E). Some interesting results are presented in this chapter.

5.2 The Solar Variability

Solar activity changes on time-scales from hours to years. Solar variability essentially originates from the interior of the Sun. Broadly one can classify the two distinct types of emissions from the Sun as (i) continuous emission and (ii) solar cycle dependent emission. This is clearly categorized in Figure 5.1.

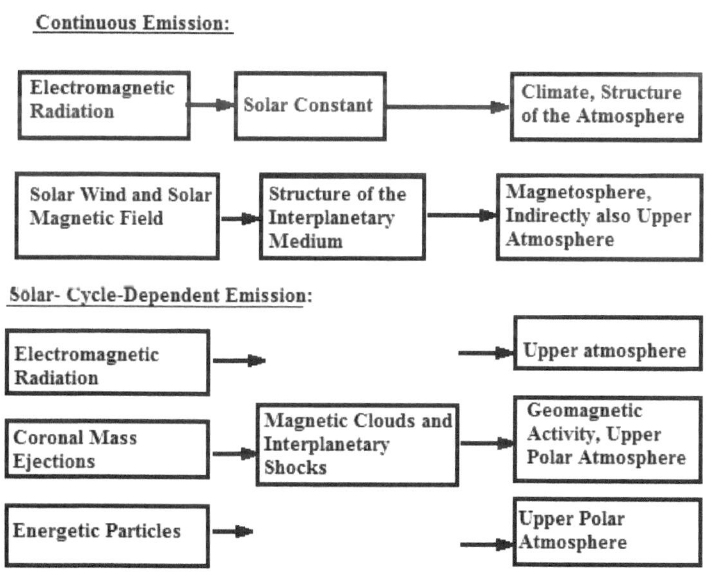

Figure 5.1 Two distinct types of emissions from the Sun

One very important aspect of solar variability is linked to solar magnetic fields generated below the convective zone in the interior. The evolution of magnetic fields is responsible for variety of manifestations such as the well-known coronal mass ejections and solar wind structures. Though the influences from the variation of the electromagnetic radiation have considered widely and largely available in literature [7-10], a changing solar wind and energetic particles have got less priority, perhaps due to its very complex scenario, which can also influence Earth either directly or indirectly through modulating the cosmic ray flux.

5.3 Extension of Solar Corona into Interplanetary Medium

The plasma corona region consists of completely ionized material containing mainly protons and electrons trapped by the overall magnetic field lines of the Sun. Figure 5.2 shows a schematic representation of the solar corona and its extension into the interplanetary medium. The figure reveals the principal features involved in the process like the exceptionally large coronal streamers. Light arises from scattering of solar radiation by free electrons in the ionized gas. The shape of the corona depends on the distribution of the magnetic field lines. It is possible to study transient features due to large scale mass ejections through the corona. The frequency range involved in the solar corona region is presented in Figure 5.3. In this figure the ranges of frequencies connect to certain portions of the solar atmosphere. The selection of the fixed frequency of 406.7 MHz of the receiver is evident as it can successfully provide our interest of reception of radio signal from the corona region of the Sun.

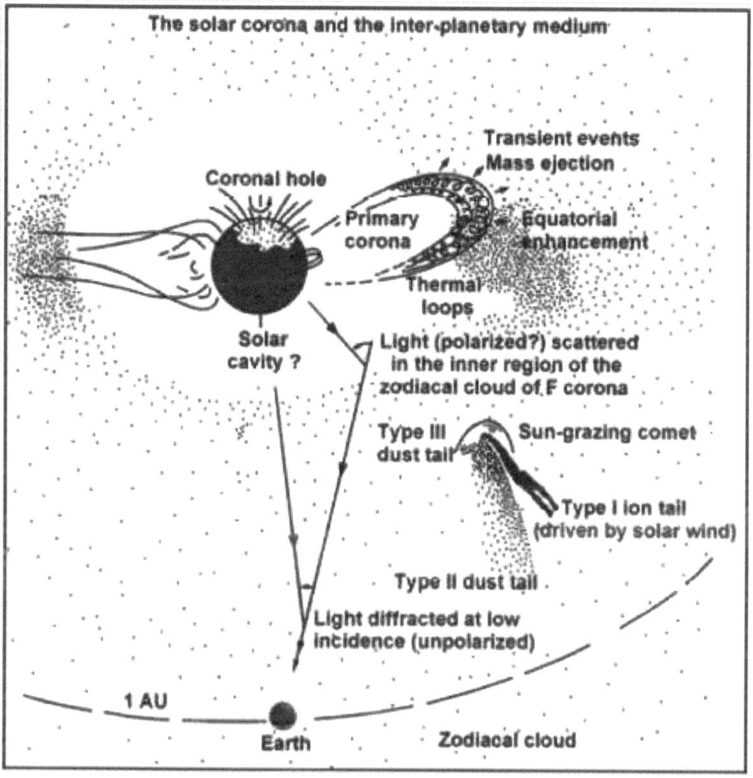

Figure 5.2 Schematic representation of the solar corona and its extensions into the interplanetary medium [1]

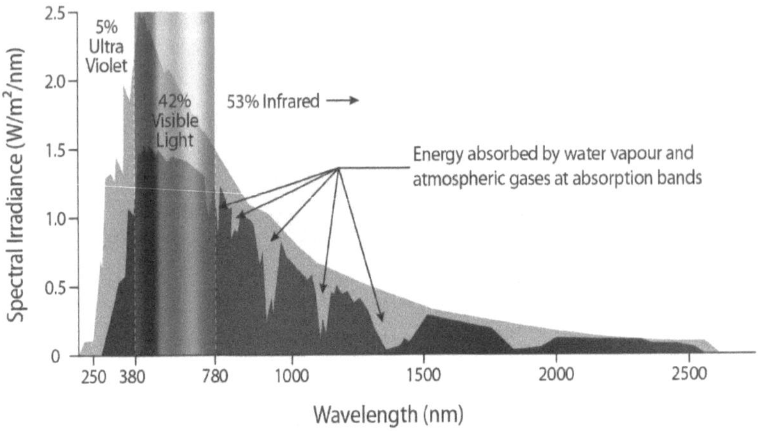

Figure 5.3 Solar Frequency Spectrums

5.4 Radio Signal Records

Two typical patterns of radio emissions originating from solar corona received at 406.7 MHz corresponding to two different dates are shown in Figure 5.4 (a) and 5.4 (b). The figure exhibits that the voltages, when the two cases are taken into account, vary significantly from 0.16 volt [in Figure 5.4 (a)] to 0.37 volt [in Figure 5.4 (b)]. Similar prominent variations we have noted in many occasions.

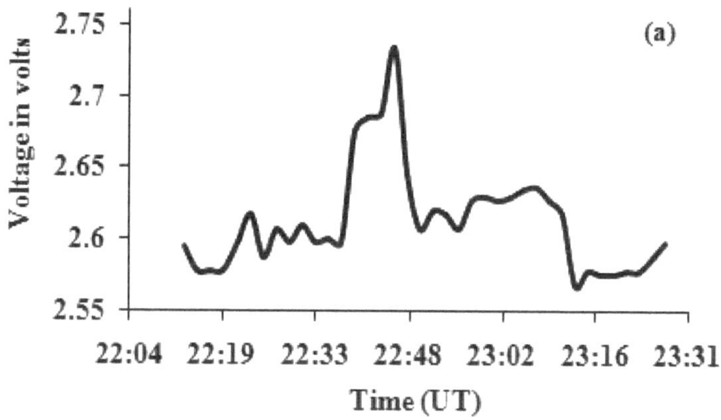

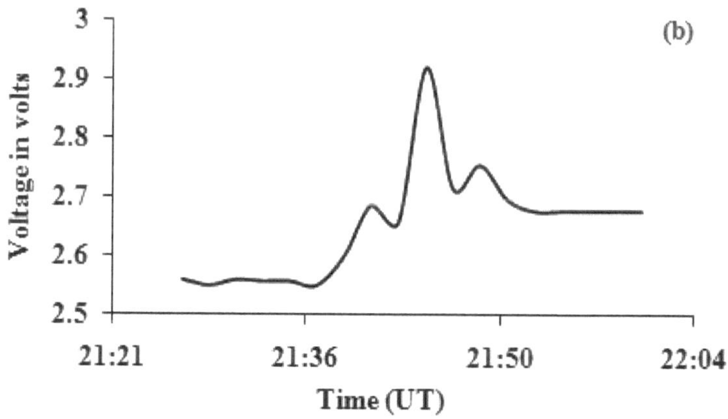

Figure 5.4 Typical variations of radio signals due to solar corona at 406.7 MHz on (a) 25-04-2011 and (b) 28-04-2011

Radio signals owing to solar corona received at different dates when analyzed have shown different characteristic patterns with a variation in the signal levels. In Figure 5.5 we have shown such characteristic differences for three different dates. These varying characteristics are not uncommon as the solar coronas are well classified into different types in shape depending on its point of origin. Out of the various type of corona, the 'Helmet streamers' are large cap-like coronal structures with long pointed peaks which usually overlie sunspots and active regions. They are formed by a network of magnetic loops that connect the sunspots in active regions. The closed magnetic field lines trap the electrically charged coronal gases for forming relatively dense structures. The pointed peaks are produced by the action of the solar wind blowing away from the Sun in the spaces between the streamers. The second category is the 'polar plumes' which are long thin streamers projecting outward from the Sun's north and south poles and are associated with the "open" magnetic field lines at the Sun's poles. These are formed by the action of the solar wind in similar way as the peaks on the helmet streamers. 'Coronal loops' are observed around sunspots and in active regions. These structures are related to the closed magnetic field lines that connect magnetic regions on the solar surface. 'Coronal holes' are regions where the corona is dark and are associated with "open" magnetic field lines. The high-speed solar wind is known to originate in coronal holes. The 'White-Light' coronas, the 'Emission Line' corona, the 'X-ray Corona' are other important types with characteristic differences. The varying nature of enhancement in the received radio signal may presumably be associated with the type of corona responsible for producing the radio signal nature.

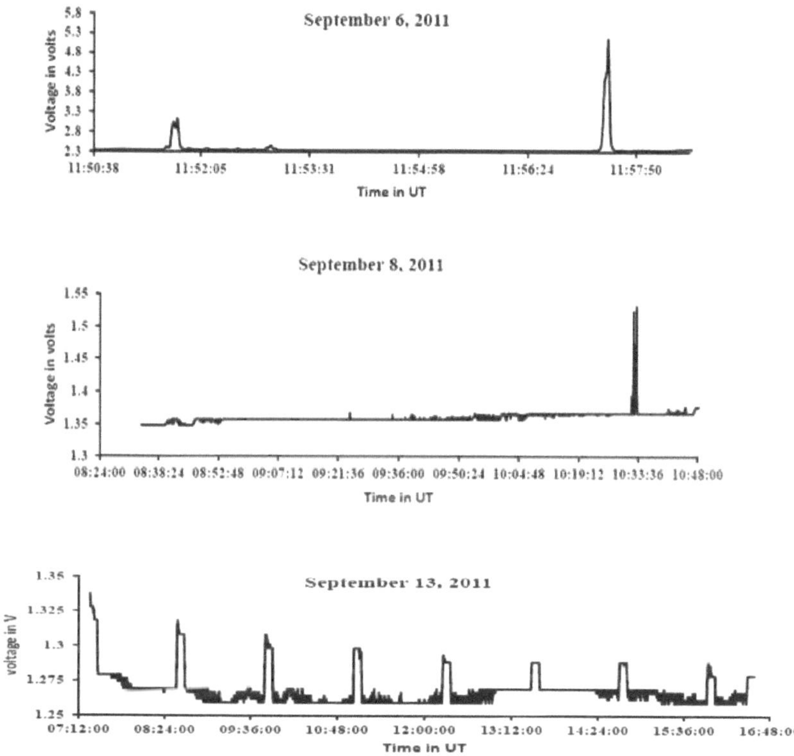

Figure 5.5 Varying characteristics of radio signal due to solar corona

5.5 Variations of Signal Strength during Solar Radio Bursts

In course of our observation, we have simultaneously received radio signals originated from solar corona by using 406.7 MHz radio telescope and the solar flare emitted from the surface of the Sun by using log periodic dipole antenna (LPDA) in the frequency range 50 to 300 MHz at Kalyani, West Bengal, India, Besides this investigation, for studying the effect of radio signal propagation we have considered alternative search strategy for interstellar communication as well as propagation of radio signals during solar eclipse [11, 12] in our laboratory. It is purpose to examine the variation of signal strength at the two systems by using two separate antennas and the associated receiving system; one operated at a particular frequency and the other operated in a small frequency band. In general, there are five type of solar radio bursts, classified as type I, type II, type III, type IV and type V. But we have received type II and type III bursts only through our receiving system. Out of the two, the type II bursts are slow frequency drift bursts, usually accompanied by a stronger intensity and lies in the frequency range of 20 MHz to 150 MHz, Flares, proton emission, magneto hydrodynamic and shock waves are often associated with type II bursts. On the other hand, the type III radio bursts are fast frequency drift bursts and can occur in groups in the frequency range of 10 kHz to 1 GHz. In the study we have also determined the effect of various geomagnetic parameters at such times covering a period of four months August 2012 to November 2012.

5.6 Results

Figure 5.6 shows some typical variations of solar bursts recorded by the two systems of observations corresponding to two specified frequencies, 406.7 MHz and 100 MHz. In the typical record we have chosen to reproduce the radio signal data of solar bursts at a fixed frequency of 100 MHz due to its prominent nature in the record. For the months of August and September prominent bursts as recorded by the two receivers are shown. The X-axes in the records represent the same time scale while the Y-axes indicate the amplitudes. It is interesting to note that the amplitude of the signals received at 406.7 MHz is much larger than that received at 100 MHz.

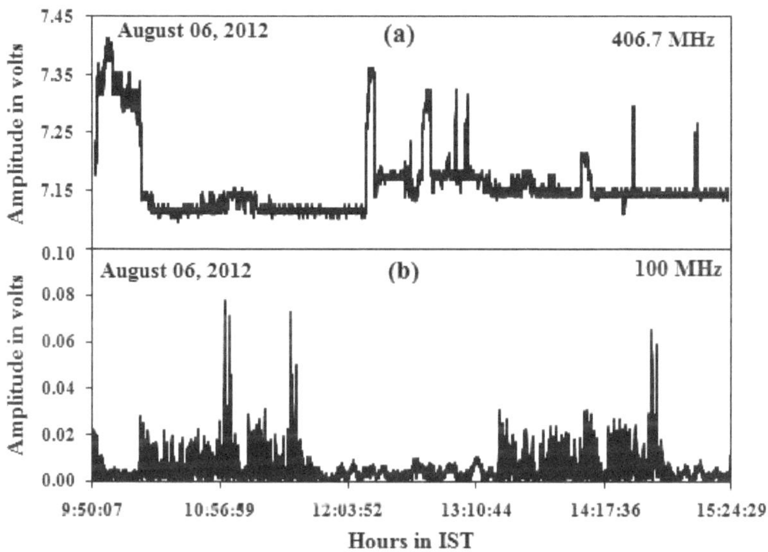

Figure 5.6 Some typical variations of solar bursts recorded at (a) 406.7 MHz, and (b) 100 MHz on August 06, 2012;

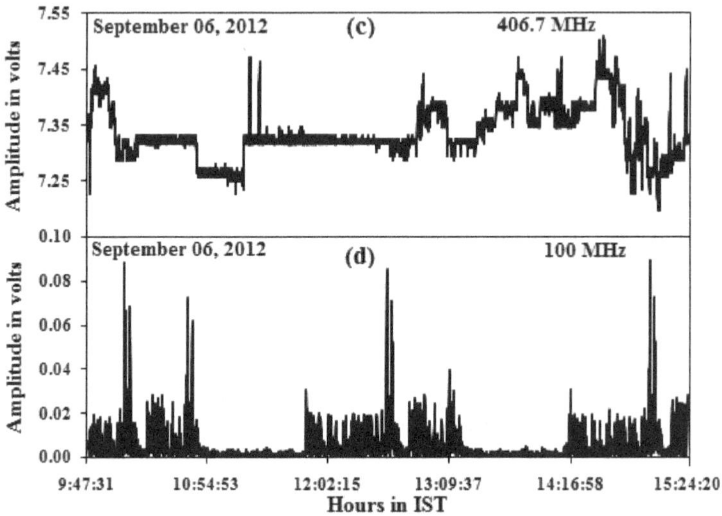

Figure 5.6 the other two samples corresponding to the said two different frequencies for September 06, 2012

5.6.1 Signal Strength Variations due to Solar Bursts

Some other records related to signal strength variations due to solar bursts have shown in Figure 5.7, which also exhibit identical behavior of the characteristic variation as pointed out in the typical variations of Figure 5.6. However, it is seen that for radio signals at 406.7 MHz there is a more prominent variations in the level. These varying characteristics are not uncommon as the solar flare itself produces variety of shapes with changing strengths depending on its positions. During our observational period using LPDA, from August 2012 to November 2012, we noted that for majority of the cases the maximum amplitude of the response lies in the frequency range from 100 MHz to 110 MHz as indicated in some others records exhibited in Figure 5.7.

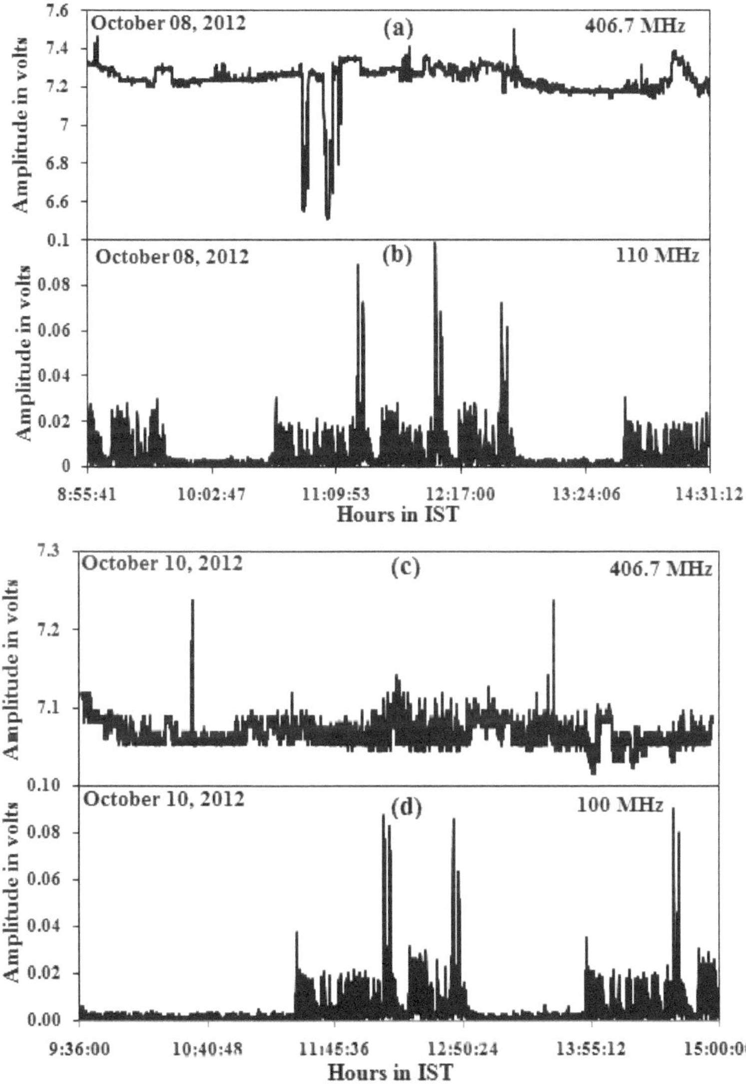

Figure 5.7 Some other records of solar bursts recorded at the two frequencies at different dates: (a) and (b) for October 08, 2012, (c) and (d) for October 10, 2012

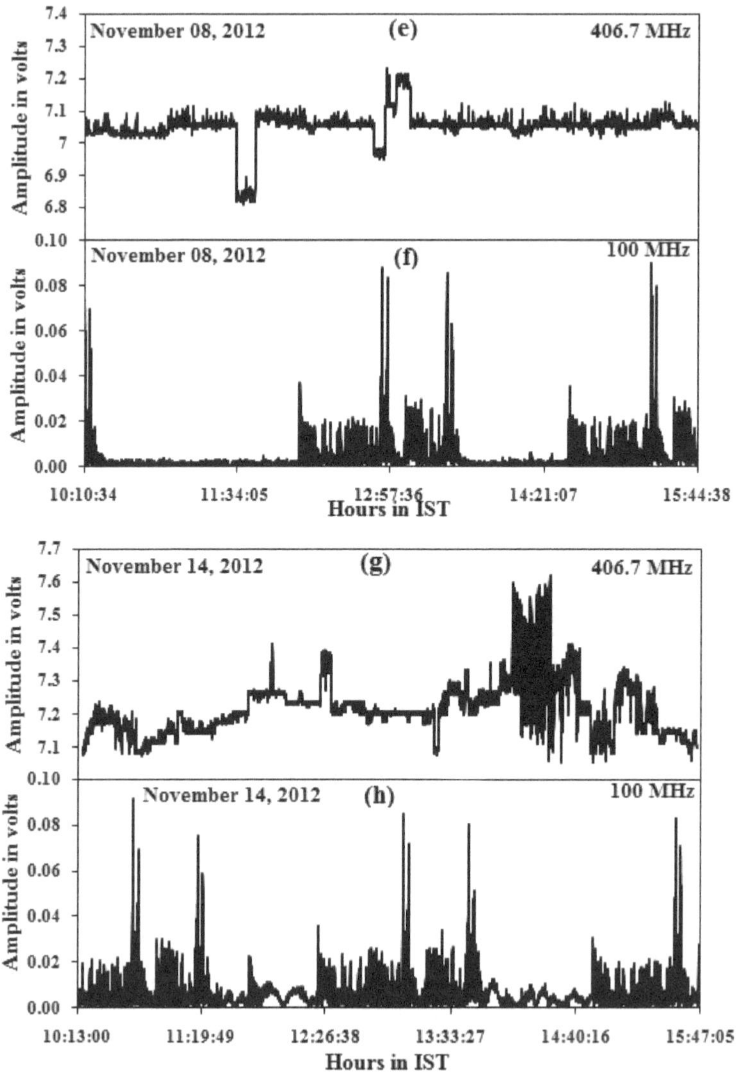

Figure 5.7 Some other records of solar bursts recorded at the two frequencies at different dates: (e) and (f) for November 08, 2012, (g) and (h) for November 14, 2012

The modern classification system for solar flares uses the letters A, B, C, M, or X, according to the peak flux in W/m2 of soft X-rays with wavelengths 0.1 to 0.8 nm as measured by GOES satellites in geosynchronous orbit which is given in table 1. This system was originally devised in 1970 and included only the letters C, M, and X [8]. These letters were chosen to avoid confusion with other optical classification systems. The A and B classes were added in the 1990s as instruments became more sensitive to weaker flares. The strength of an event within a class is noted by a numerical suffix ranging from 1 up to, but excluding, 10, which is also the factor for that event within the class. Hence, an X2 flare is twice the strength of an X1 flare, an X3 flare is three times as powerful as an X1. M-class flares are a tenth the size of X-class flares with the same numeric suffix. An X2 is four times more powerful than an M5 flare. X-class flares with a peak flux that exceeds 10−3 W/m2 may be noted with a numerical suffix equal to or greater than 10 [9]. An earlier classification system, sometimes referred to as the flare importance, was based on H-α spectral observations is given in table 2 [10]. The scheme uses both the intensity and emitting surface. The classification in intensity is qualitative, referring to the flares as: faint (f), normal (n), or brilliant (b). The emitting surface is measured in terms of millionths of the hemisphere and is described below.

The space weather monitor data graph below (Figure 5.8) shows the changes to the signal strength caused by sunrise, sunset, and solar flares. The solar flares are classified by strength, from low to high as A, B, C, M and X.

Table 5.1 Classification of the solar flares according to the intensity [8]

Classification	Peak flux range (W/m^2)
A	< 10^{-7}
B	$10^{-7} - 10^{-6}$
C	$10^{-6} - 10^{-5}$
M	$10^{-5} - 10^{-4}$
X	> 10^{-4}

Table 5.2 Earlier classification system of the solar flares based on H-alpha spectral observations [10]

Classification	Corrected area (millionths of hemisphere)
S	< 100
1	100–250
2	250–600
3	600–1200
4	> 1200

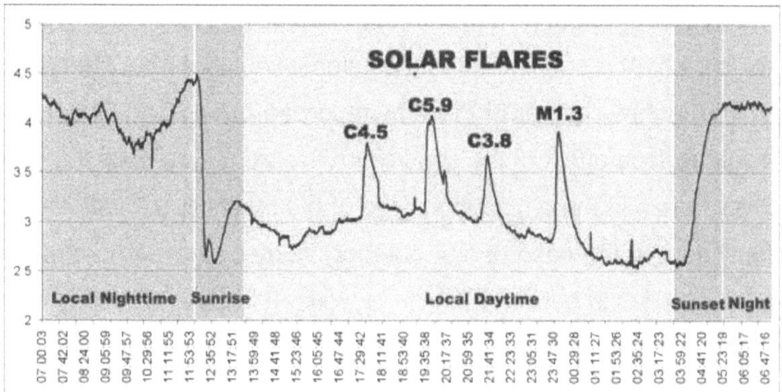

Figure 5.8 The space weather monitor data plot shows the variations in signal strength due to sunrise, sunset, and 4 solar flares classified as A, B, C, M, and X.

The solar flares we have received when compared with the NASA reported time exhibit a good time coincidence. The solar radio burst, associated with solar flares, we have received when compared with the NASA reported time exhibit a good time coincidence. It may be noted that the flares we have analyzed are mostly of M-type solar category. The signal strengths of the coronal effect (electromagnetic radiation from the solar corona, we call it coronal effect) at such times were very high compared to the surface effect (that radiation emitted from the sun, we call it surface effect) of the Sun due to the tremendous temperature of coronal area belonged to the outer most layer of the solar atmosphere. Actually, the temperature Is the vital part of solar radiation and it controlled the radiation strength. According to Wien's law the temperature increases with the decrease of wavelength of the radiation *i.e.* the temperature is increased with the increase of frequency of the radiation. So, the energy of the radiation becomes high with the higher value of the frequency. This explains why the amplitude of these signals emitted from the corona is much higher compared to other recorded burst data. The signal level of surface effect of the sun lies on the 'zero' line while the peak activity varies in the amplitude range of 0.06 volts to 0.1 volt. On the other hand, in the case of first category related to solar corona, the signal level lies in between 7.0 to 7.2 volts. It is also noted that there are some time differences (There are small times differences between the peaks of the two signals receive by us through two receivers) occasionally between these two types of emissions. A list of solar flares related to Figure 5.6 and Figure 5.7 is presented in Table 5.3. The corresponding time (in IST) and the sunspot region have also shown.

Table 5.3 The list of prominent solar flare observed during August to November in 2012

Date (Year 2012)	Time (IST)	Flare Type	Sunspot Region
August 06	10:08	M1.6	1542
September 06	9:43	M1.6	1560
October 08	16:47	M2.3	Un-numbered eastern limb
October 10	10:34	M1.0	Un-numbered eastern limb
November 08	7:53	M1.7	1611
November 14	9:34	M1.1	1613

5.7 Solar Wind Parameters

The solar wind is constituted by highly energetic charged particles. This is mainly composed of electrons and protons, released from the upper atmosphere of the Sun. The temperature and the speed of these particles are changed from time to time. These particles with huge amount of kinetic energy enable them to escape from the solar surface. Corona is the determining factor for the source of solar wind. The temperature of the corona being extremely high, the solar gravitational force cannot hold these particles. With a view to examine the contribution of the solar wind parameters, we have collected the solar wind data (Bartol Research Institute, http://umtof.umd.edu/pm/crn/) for the concerned period.

In Figure 5.8 we have plotted the solar wind velocity (proton velocity), proton density, most probable velocity and their arrival direction for the four months. It is highly interesting to see from Figure 5.8 that both the solar wind velocity and proton density are increased in the dates of solar activity producing severe bursts. We have also noted that there are sudden disturbances in the arrival direction of solar wind particles. These increments are assumed to happen due to emitted solar flare from the outer surface of the sun. The solar wind particles are coming from the sun. They are arrived to the earth atmosphere from north or from south. Here positive sign indicates they are coming from south while negative sign specifies the incoming particles are coming from the north side.

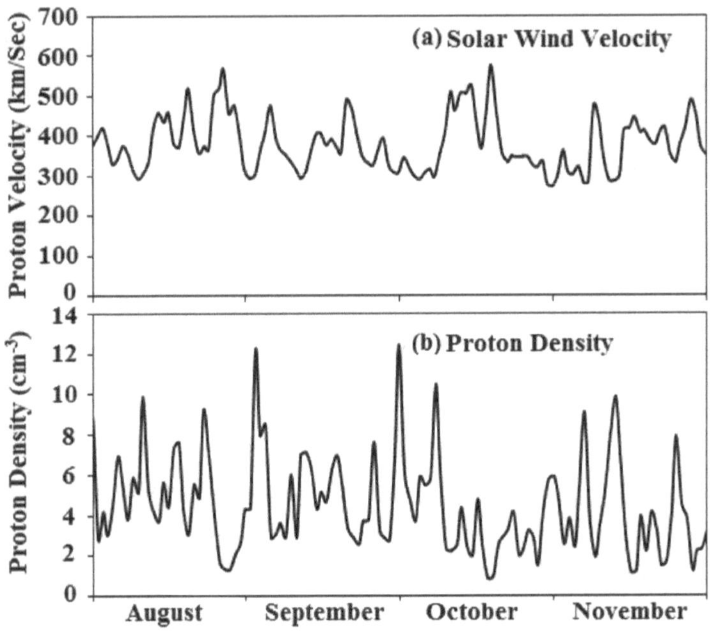

Figure 5.8 The solar wind parameter in the month of August to November 2012: (a) corresponds to solar wind velocity, (b) proton density,

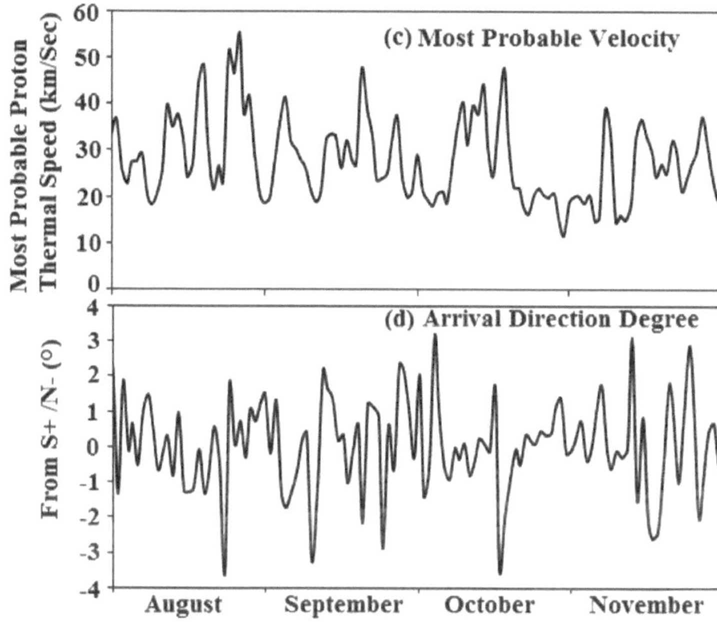

Figure 5.8 The solar wind parameter in the month of August to November 2012. (c) most probable velocity and (d) arrival direction of solar wind particles

5.8 Disturbance Storm Time Index

The disturbance storm time index (DST) is a geomagnetic index, which checks worldwide magnetic storm levels. Actually, this DST index is a measure of geomagnetic activity used to find the severity of magnetic storms. It is based on the average value of the horizontal component of the Earth's magnetic field measured hourly at four near-equatorial geomagnetic observatories [13]. In the case of a classic magnetic storm, the DST exhibits a sudden rise, corresponding to the sudden commencement of the storm and then decreases sharply as the ring current intensifies. In our analysis, we have plotted the DST index in the month of August, September, October and November 2012. It should be noted that there has some sudden fall in the DST strength in August 06, September 06, October 08, October 10, November 08 and November 14, 2012. This happens due to the interaction of solar radiation with the magnetic field of the Earth. It appears from Figure 5.9 that the amplitude is dropped from the negative value of 32 nT on August 06, 2012 to the negative value of 110 nT on November 14, 2012.

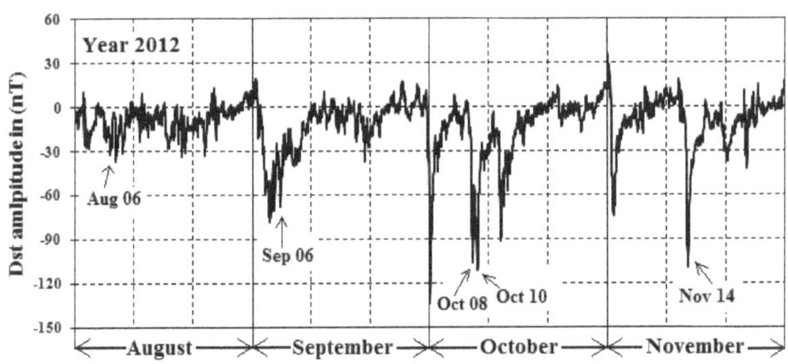

Figure 5.9 Disturbance time storm index in the month of August to November 2012

5.9 Other Geomagnetic Indices

The K-index quantifies disturbances in the horizontal component of earth's magnetic field with an integer in the range 0-9 with 1 being calm and 5 or more indicating a geomagnetic storm. It is derived from the maximum fluctuations of horizontal components observed on a magnetometer during a three-hour interval. The label

'K' comes from the German word 'Kennziffer' meaning 'characteristic digit.' The A_p index is a measure of the general level of geomagnetic activity over the globe for a given day and is derived from measurements made at a number of stations world-wide of the variation of the geomagnetic field due to currents flowing in the earth's ionosphere and magnetosphere. C index is a subjective daily character figure of geomagnetic activity for a single observatory for Earth and considered 0 – for very quiet magnetic condition, 1 – for moderately disturb condition and 2 – for severely disturb condition. C_p index is observed from sum of eight daily values of c_p index range from 0 to 2.5 wherein 2.5 is taken as most disturbed. The *aa*-index is a simple global geomagnetic activity index and is derived from the K indices and has units of nT.

It is seen from Figure 5.10 that the magnitudes of K_p, C_p, Ap and *aa* are increased between September 01, 2012 to September 09, 2012 showing peak value around September 06, 2012. Again, a second peak is noticed during October 06, 2012 to October 12, 2012 when the received solar bursts become very prominent.

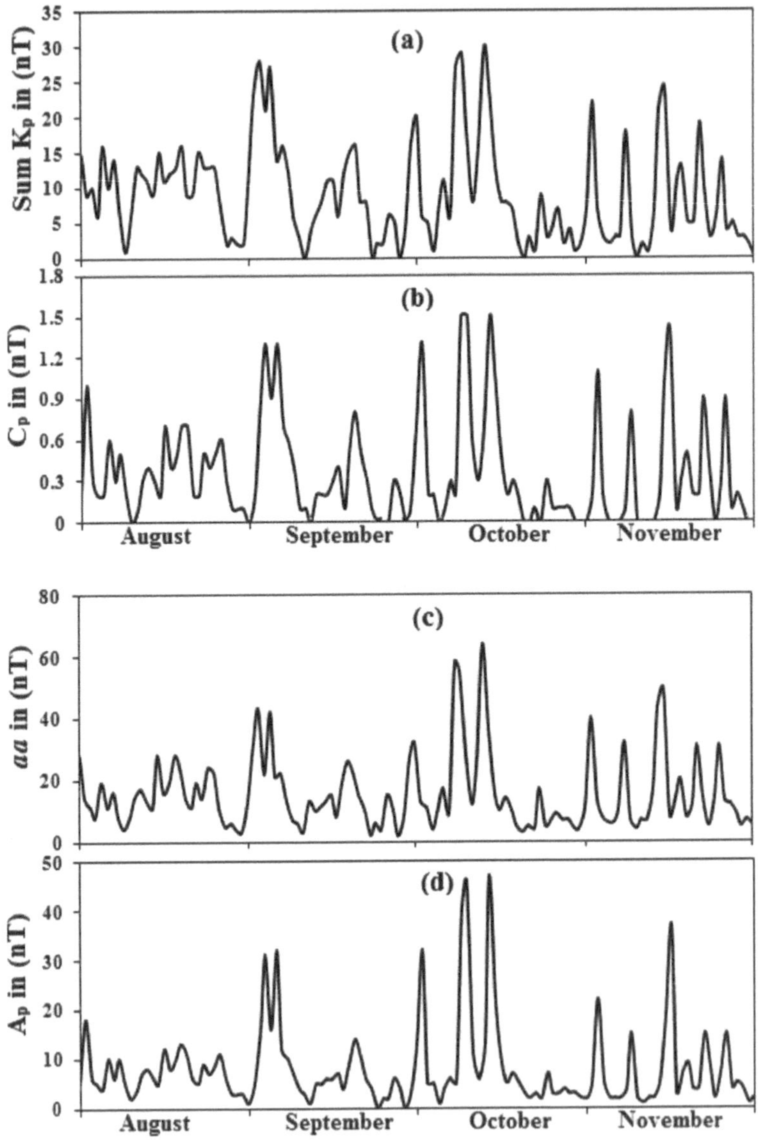

Figure 5.10 The Solar Geomagnetic indices (Sum K_p, C_p, *aa* and A_p) for the corresponding months

5.9 Discussion

Observations of the solar corona in front of the solar disk have established a direct relationship between coronal features and solar wind. Regions of solar atmospheres known as coronal holes closely correspond to the high-speed component of the solar wind. The solar wind is able to generate significant heating of the lower thermosphere at high latitudes by direct particle precipitation and also by generating upper atmospheric ionization that may influence the global electric circuit. Due to coupling with the magnetosphere the solar wind drives ionospheric currents at high latitudes which in turn accelerate the neutral atmosphere. Some investigations [14, 15] have suggested a strong relationship between the electric field of the solar wind and a pressure phenomenon in the north Atlantic called as the North Atlantic Oscillation (NAO) [16]. A substantial portion of the climate variability in the Atlantic sector is highly related to the NAO with variations occurring on a wide range of scales. Some model studies [17, 18] indicate that the influence of geomagnetic activity on the stratosphere is comparable to that of ultraviolet flux variation between solar minimum and solar maximum. Bucha and Bucha [19] observed an intensification of both thermo-spheric and tropospheric flows associated with strong geomagnetic activity and suggested a mechanism where downward winds are generated in the polar cap of the thermosphere and penetrate to the stratosphere and troposphere [20-22].

Early observations of the visible spectrum of the corona showed bright emission lines at wavelengths that did not correspond to any known materials. From this, astronomers were able to propose the existence of "coronium" as the principal gas in the corona. However, the true nature of the corona remained a mystery until it was determined that the coronal gases are super-heated to temperatures greater than 1,000,000°C. At these high temperatures the two dominant elements hydrogen and helium are completely stripped of their electrons. Even the minor elements like carbon, nitrogen, and oxygen are stripped down to bare nuclei. In fact, in this intense heat only the heavier trace elements like iron and calcium are able to retain a few of their electrons [16]. The emission from these highly ionized elements produces the spectral emission lines that were mysterious to early astronomers. It is by now possible to produce artificial eclipses in coronagraphs that cover the disk of the Sun and filter out everything except the emission due to these coronal ions. These coronagraphs produce images of the "emission line corona.

The fluctuations, which are usually independent for each forcing parameter, e.g., Forbush decreases, solar flares, ground level events, etc., provide a possible way for distinguishing between the different processes [24-28]. However, the solar processes being highly variable at time-scales of hours to days, the task of identifying a solar signal in the climate data can mask the solar effect and there may be a strong impact on the data analysis and subsequently to establish a possible solar terrestrial relationship.

The corona is basically a part of the solar flare (Coronal mass ejections are often related with other forms of solar activity, most notably solar flares, although a causal relationship has not been established). The temperature of the corona is so high that it may throw some charged particles as solar wind. These may build up the electric field that again accelerates the charged particles [29]. The relation of solar wind and coronal feature can be established by observing the solar corona in front of solar disk. Coronal holes closely communicate with high-speed component of solar wind and these solar wind particles can interact with upper atmosphere and may influence the global electric circuit at high latitude [30]. The signal strengths of the coronal effect in our observed data were very high compared to the surface effect of the Sun. In fact, the signal level of surface effect of the Sun at the peak activity appears to vary in amplitude from 0.06 volts to 0.1 volt while those related to solar corona exhibit a change in between 7.0 to 7.2 volts. The solar wind also couples with magnetosphere and hence derives the ionosphere so that it again interacts with neutral atmosphere [14, 31]. These solar wind particles, i. e. the so-called charged particles interact with Earth's magnetic field and influence the geomagnetic indices. Due to the active role of the charged particles, the geomagnetic indices are enhanced at the time of solar bursts when the DST index reduces rapidly.

References

1. A. B. Bhattacharya, R. Bhattacharya, S. Das, D. Halder, A. Sarkar, A. Bhoumick, T. Das, and M. Debnath, "Implementation of Fast Fourier Transform for detecting Solar Bursts", *Research Journal of Earth and Planetary Studies*, 1, 009-018, 2011.

2. R. Behlke, IRF Scientific Report 275, Swedish Institute of Space Physics, Sweden, 2001.

3. K. Shibata and T. Magara, "Solar Flares: Magneto-hydrodynamic Processes", *Living Reviews in Solar Physics*, 8, 6, 2011.

4. A. B. Bhattacharya, D. Halder, A. Sarkar, J. Pandit and B. Raha, "Studying transient features through the solar corona by 406.7 MHz Radio telescope", *International Journal of Multidisciplinary Research and Advances in Engineering*, 4, 225-234, 2012.

5. A. B. Bhattacharya, S. Joardar and R. Bhattacharya, *Astronomy and Astrophysics*. Infinity Science Press, Massachusetts, USA, 2008.

6. V. Williams, J. Austin and J. D. Haigh, "Model simulations of the impact of the 27-day solar Rotation period on stratospheric ozone and temperature", *Advances in Space Research*. 27, 1933–1942, 2001.

7. J. D. Richardson, C. Wang and K. I. Paularena, "The solar wind: from solar minimum to solar maximum", *Advances in Space Research*, 27, 471-479, 2001.

8. K. Kudela, J. Rybak, A. Anatalova and M. Storini, "Time evolution of low-frequency periodicities in cosmic ray intensity", *Solar Physics*, 2005, 165-175, 2002.

9. K. Labitzke and K. Matthes, "Eleven-year solar cycle variations in the atmosphere: observations, mechanisms and models", *Holocene*, 13, 311-317, 2003.

10. D. Shindell, D. Rind, N. Balachandran, J. Lean and P. Lonergan, "Solar cycle variability, ozone, and climate", *Science*, 284, 305-308, 1999.

11. A. B. Bhattacharya, S. Sarkar and R. Bhattacharya, "An alternative search strategy for interstellar communication", *Indian Journal of Physics*, 84, 511-515, 2010.

12. R. Bhattacharya, A. Nag, R. Guha, A. Bhoumick, S. De and A. B. Bhattacharya, "Solar Eclipse Effects of 22 July 2009 on the Propagation of Radio Signals" *Indian Journal of Physics*, 84, 1587-1595, 2010.

13. N. O. Bakare and V. U. Chukwuma, "Relationship between Dst and solar wind conditions during intense

geomagnetic storms", *Indian journal of Radio and Space Physics*, 39, 150-155, 2010.

14. F. Boberg, and H. Lundstedt, "Solar wind electric field modulation of the NAO: A correlation analysis in the lower atmosphere", *Geophysical Research Letters,* 30, 1825, 2003.

15. F. Boberg, and H. Lundstedt, "Solar wind variations related to fluctuations of the North Atlantic Oscillation", *Geophysical Research Letters,* 29, 1718, 2002.

16. J. Marshall, Y. Kushnir, D. Battisti, P. Changd, A. Czaja, R. Dickson, J. Hurrell, M. Mccartney, R. Saravanan and M. Visbeck, "North Atlantic climate variability: phenomena, impacts and mechanisms Impacts and Mechanisms International Journal of Climatology", *International Journal of Climatology*, 21, 1863–1898, 2001.

17. N. F. Arnold and T. R. Robinson, "Solar magnetic flux influences on the dynamics of the winter middle atmosphere", *Geophysical Research Letters,* 28, 2381–2384, 2001.

18. N. F. Arnold and T. R. Robinson, "Solar cycle changes to planetary wave propagation and their influence on the middle atmosphere circulation", *Annals of Geophysics*, 16, 69-76, 1998.

19. V. Bucha and V. Bucha "Geomagnetic forcing of changes in climate and in the atmospheric circulation", *Journal of Atmospherics and Solar Terrestrial Physics*, 60, 145-169, 1998.

20. R. W. Fairbridge, *Encyclopedia of Atmospheric Sciences and Astrogeology*, Encyclopedia of Earth Sciences Series, 2, 1967.

21. P. Zhang, A. D. Gilbert and K. Zhang "Nonlinear dynamo action in rotating convection and shear", *Journal of Fluid Mechanics*, 546, 25–49, 2006.

22. R. Muscheler, J. Beer and B. Kromer, "Long-term Climate Variations and Solar Effects", *In Proc., ISCS 2003 symposium,* Solar Variability as an Input to Earth's Environment, Tatranska Lomnica, Slovakia, 23-28 June 2003, ESA SP-535, September 2003.

23. A. B. Bhattacharya, *Search for Extraterrestrial Intelligence through Radio Astronomical Studies*, Lambert Academic Publishing, Germany, 2012.

24. H. Svensmark, "Influence of cosmic rays on Earth's climate", *Physical Review Letters*, 81, 5027-5030, 1998.

25. G. C. Reid, "Solar Total Irradiance Variations and the Global Sea-Surface Temperature Record", *Journal of Geophysical Research-Atmospheres*, 96, 2835-2844, 1991.

26. E. Friis-Christensen and K. Lassen, "Length of the Solar-Cycle - An Indicator of Solar-Activity Closely Associated with Climate", *Science*, 254, 698-700, 1991.

27. J. M. Mitchell, C. W. Stockton and D. M. Meko, "Evidence of a 22-year rhythm of drought in the western United States related to the hale solar cycle since the 17th century, in Solar-Terrestrial Influences on Weather and Climate". Editor B. M. McCormac and T. A. Seliga, D., Reidel publishing Company, 1979.

28. D. Moss and D. Sokoloff, "Dynamo generated toroidal magnetic fields in rapidly rotating stars", *Astronomy & Astrophysics*, 497, 829–833, 2009.

29. U. Feldman, E. Landi and N. A. Schwadron, "On the sources of fast and slow solar wind", *Journal of Geophysical Research,* 110, A07109.1–A07109.12, 2005.

30. B. A. Tinsley, "Influence of solar wind on the global electric circuit and inferred effects on cloud microphysics, temperature and dynamics in the troposphere", *Space Science Reviews*, 94, 231-258, 2000.

31. D. F. Webb and T. A. Howard, "Coronal Mass Ejections: Observations", *Living Reviews in Solar Physics*, 9, 3-83, 2012.

Chapter 6

Investigation on VLF Radio Signals & Some Interrelated Parameters

6.1 VLF Signals from Various Natural Phenomena

Very Low Frequency (VLF) radio signals generally refer to the range 3 kHz to 30 kHz. On our planet, the main effective source of VLF wave is lightning, though there are some other sources too, which are Quasi-electrostatic whistler mode wave, solar activity, lightning energy, electron precipitation from the radiation belts, cosmic gamma-rays, earthquake, meteor shower (e. g. Orionids) near earth's surface etc. VLF wave reflect at the D region of earth's ionosphere level and as a result it efficiently guided in the Earth-ionosphere waveguide to global distances. This VLF signals interact with earth's atmosphere as well as it penetrates into seawater [1, 2]. It is noticed that in various natural phenomenon, like earthquake, meteor shower, a planet or asteroid is coming close to the earth's atmosphere, exists radio signal which can interact with earth's D-region of ionosphere and that why VLF signals are reflect. Some few phenomena are indicated here with.

6.2 VLF Atmospherics Record

A study of the influence of the monsoon on the atmospherics level has been a subject of great interest for the last few decades and many scattered reports have been published in the literature from time to time [3, 4]. Close association of atmospherics during both the pre monsoon and monsoon thunderstorms have been published by different authors from observations taken at different parts of the globe [5-9] including observations in the VLF band [10]. Figure 6.1 exhibits the characteristic variations in the atmospherics record at 40 kHz as recorded on June 9, 2011 when the monsoon was in formation in the Bay of Bengal. After a distinct and a very short period fall of the intensity level it reveals an enhancement and after a clear pause time there is a sudden fall of atmospherics (SFA) which subsequently regains to the normal level. This characteristic variation is found to be associated with the thundershower at such times as confirmed also from the reports of the meteorological departments.

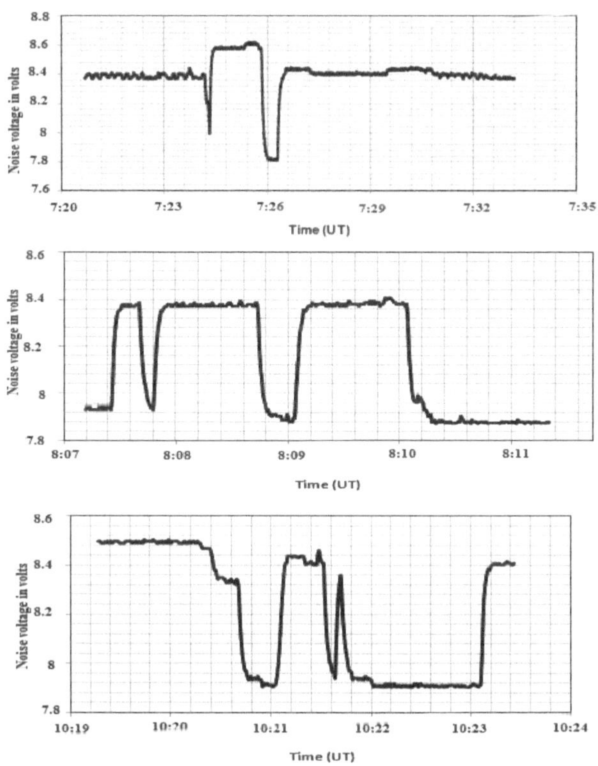

Figure 6.1 precursory effects in the atmospherics level before the arrival of South Bengal monsoon. The data were recorded on June 9, 2011

The enhancement and the pause time in the intensity level was associated with the mature stage of the thundercloud while the SFA effect was closely related to short period sharp rainfall as also visually noted over the observatory. This precursory effect in the atmospherics level before the arrival of the South Bengal monsoon appears to be closely associated with the onset of the tropical monsoon under prevalent atmospheric conditions. The characteristic variation in the atmospheric level, a few days before the onset of local monsoon, is a typical characteristic we have recorded in our laboratory and may be taken as an interesting precursory effect not reported elsewhere before which might be due to lack of careful investigation emphasizing the problem of interest considered here. Figure 6.2, on the other hand, shows a typical pattern we have noted in our records when the monsoon gets a very stable shape locally and overcast thunderclouds repeatedly contribute through lightning discharges. Following sudden enhancement and subsequent pause, recovery begins in the level of atmospherics and the field intensity comes back almost to its former level. A close scrutiny of the records presented in Figure 6.1 and Figure 6.2 reveals interesting features. Some Dominant features of atmospherics either in the form of sudden enhancement or sudden fall related to the typical transitions in the atmospherics level are presented in Table 6.1. Figure 6.3 exhibits some other typical atmospherics record at 40 kHz when the monsoon was converted to incessant and torrential rainfall with repeated lightning discharges caused by surrounding monsoon thunderclouds. The typical records shown in the figure was received in our laboratory when the depression was located at the south-west of Kalyani on June 16, 2011.

Table 6.1 Duration and magnitude of the typical variations at 40 KHz atmospherics (calculated from Figures 6.1 and Figure 6.2)

Date of occurrence	Dominant features of atmospherics	Onset time in U.T.	Horizontal duration	Amplitude variation (V)
June 09, 2011	SFA	7:24:21	90 sec	0.20
		7:25:51	18 sec	0.60
		8:07:26	12 sec	0.46
		8:07:48	46 sec	0.48
		8:09:02	54 sec	0.49
		10:20:16	16 sec	0.58
		10:21:30	78 sec	0.56
June 15, 2011	Enhancement	8:50:00	13 min	0.60
		9:10:00	12 min	0.75
		9:38:00	12 min	0.55
		12:57:20	36 sec	0.12
		13:00:54	40 sec	0.18
		13:04:32	38 sec	0.20
		13:06:12	38 sec	0.20
		13:08:30	50 sec	0.18
		13:10:46	50 sec	0.12

		13:12:48	52 sec	0.18
		14:26:36	45 sec	0.20
		14:31:24	40 sec	0.14
		14:34:35	42 sec	0.15
		14:36:24	36 sec	0.22

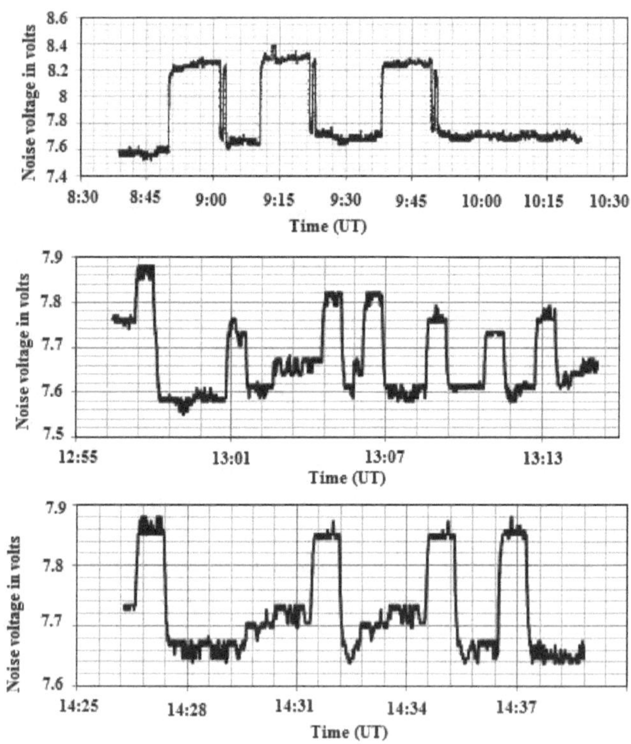

Figure 6.2 A typical patterns of atmospherics when the monsoon gets a very stable shape. The data were recorded on June 15, 2011

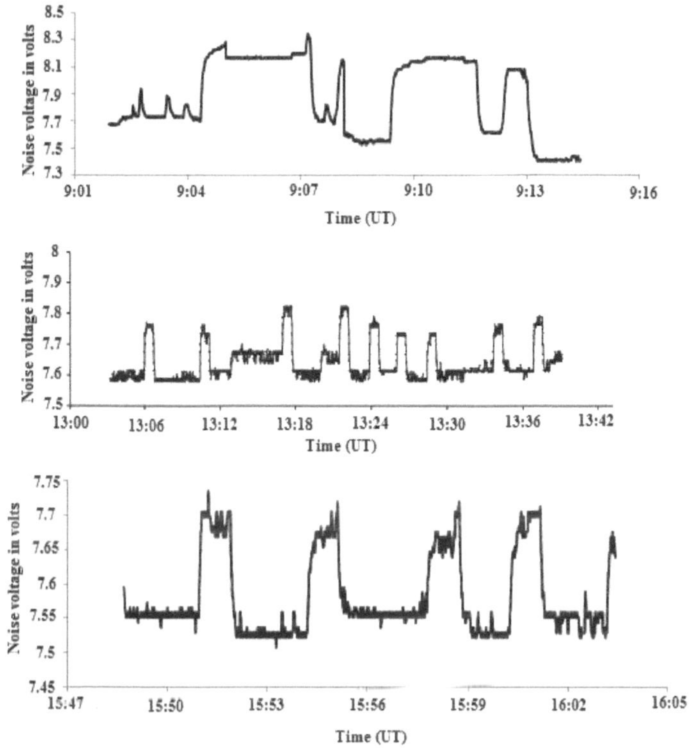

Figure 6.3 Some other typical atmospherics records during severe monsoon. The data were recorded on June 16, 2011

6.2.1 VLF Signals due to Indonesia Earthquake

VLF atmospherics at 27 kHz operating round-the-clock at Kalyani West Bengal, reveal typical variations in the noise level during the violent earthquake of magnitude 8.6 on April 11, 2012, first experienced at Indonesia and subsequently at different cities of India. In association with the earthquake followed by major cities of India, we have analyzed effect on VLF atmospherics at 27 kHz. Figure 6.4 represents the characteristic variation in the atmospheric records at Kalyani University campus.

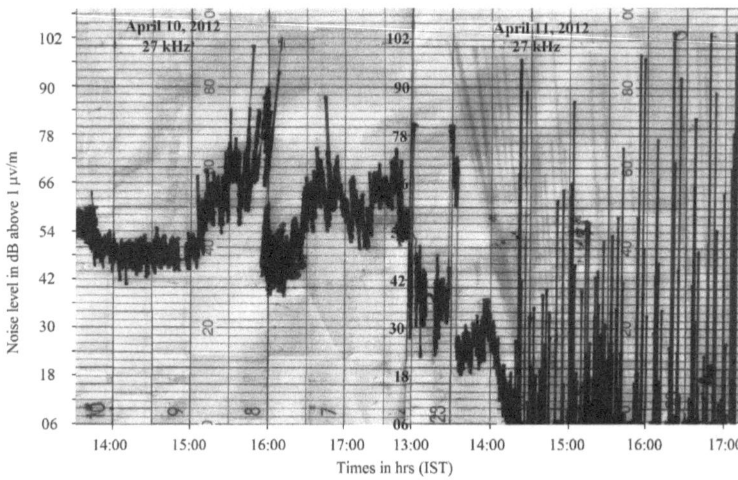

Figure 6.4 Typical records of VLF atmospherics at 27 kHz. The left-hand side record is related to local nor'wester while the right-hand side record is obtained during the violent earthquake (the ordinate is in dB above 1 μV/m)

The left-hand side record of atmospherics exhibits the typical variation in the noise level associated with a local nor'wester. The record shows a sudden enhancement of the noise level which is then repeated at higher Y-scale on April 10, 2012. Under this continued disturbed condition of the atmosphere on April 11, 2012 the violent earthquake was experienced at Indonesia and subsequently at different parts of India. Over Kolkata and its adjoining areas like Kalyani the shock was experienced at 1414 IST which was later repeated once again with relatively weak strength. It appears that in the present case, the earthquake produces induced gravity wave perturbations [11, 12] in the ionospheric D-region during daytime producing the most irregular and asymmetric fluctuation in the noise level after the event [13, 14]. It is seemed that even when the violent earthquake occurred at a sufficiently remote location from Kalyani the severe disturbances may be responsible to influence our lower ionospheric layers and may even create problems in communication system temporarily.

6.2.2 VLF Signals due to Transit of Venus across the Sun at Kalyani Sky

The transit (passage of a planet) across the disk of the Sun may be considered as a special type of eclipse. From Earth, only transits of the inner planets Mercury and Venus are possible to observe. Transits of Venus happen usually in pairs with eight years separating the two events [15-18]. We were able to observe the transits of Venus at the Kalyani sky in West Bengal on June 6, 2012 and also simultaneously recorded 40 kHz atmospherics to examine any characteristic changes at the time of such transits. Transit of Venus across the disk of the Sun reveals that there are a number of short-period fading in the atmospherics level during the period of transits of Venus. The radio signal received at 40 kHz exhibits a rapid variation of the voltage level with a sudden enhancement before the greatest position of the Venus. Observations were aimed at exploring the changes of ionization of the earth's upper atmosphere, which is known to support ionospheric propagation of radio signals. Record of the integrated field intensity of atmospherics at 40 kHz represents the variation of atmospherics during the transit events of Venus. The typical variation is presented in Figure 6.5 wherein at the background the position of the Venus across the solar disk is shown.

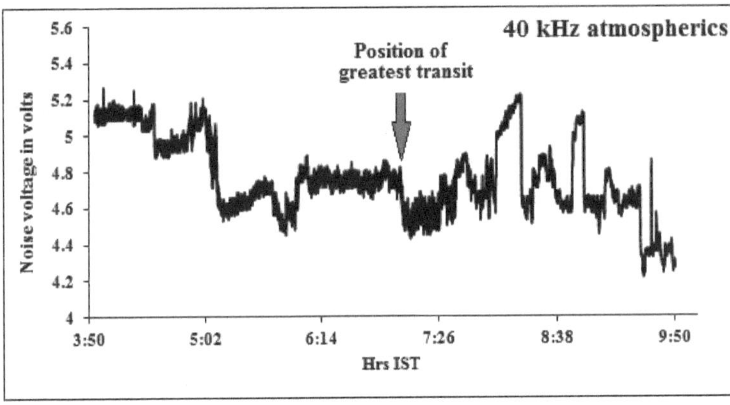

Figure 6.5 Atmospherics at 40 kHz showing reduction of the noise voltage with characteristic variations during the transits of Venus.

By a close scrutiny of the atmospherics record, it is seen that there is a distinct fall in the atmospherics level prior to the ingress interior (contact II) of the Venus and the atmospherics maintain this reduced level during the whole commencement. The figure also reveals that there are a number of short-period fading in the atmospherics level prior to the position of greatest transit. Large numbers of fade-outs with comparatively high duration were also produced in the atmospherics level after the greatest transit. When we examine 40 kHz atmospherics data, we find again sawtooth fadeouts (SF) of much shorter duration but with greater number of occurrences. Inverted saw-tooth type patterns have also been noted in some occasions. We have made a critical analysis of our VLF records during October 17 to October 27, 2011 when most of the Orionids were observed. Typical day and nighttime events have shown in Figure 6.6.

6.2.3 VLF Atmospherics Characteristics due to Orionid Meteor Shower

It is well known that a meteor can be a source of ionization in the upper atmosphere of the earth [19, 20]. The ionization in meteor burst is often harnessed for developing a communication link in the transmission, unaffected by sudden ionospheric disturbances [21-23]. The same ionization may be generally responsible at different frequency bands during propagation of signals through it [24]. An observed association of sawtooth fadeout (SF) effects in the field strength of 40 kHz VLF atmospherics with Orionid meteor showers of October 2011 is reported. A significant geomagnetic activity during major meteor showers indicates a correlation between the Orionid meteor and the magnetic phenomena. Our analysis suggests that highly ionized debris of meteors and absence of photo-detachment process are responsible for attracting meteor effects on radio signals more at night than that at daytime.

6.2.4 VLF Atmospherics Characteristics due to an Asteroid

On February 15, 2013, the asteroid 2012 DA14, approximately 45 meters in diameter, passed by our planet at a remarkably close distance. It was only about 27,700 kilometers above Earth's surface at the time of closest approach on February 15, 2013. The distance is well outside the atmosphere of the earth, but it is inside the belt of satellites in geostationary orbit located 35,800 kilometers above the earth's surface [25-31]. Figure 6.7 reveals the variation of noise voltage at 40 kHz atmospherics on 15.02.2013 in the time range when asteroid 2012 DA14 was inside the belt of satellites in geostationary orbit.

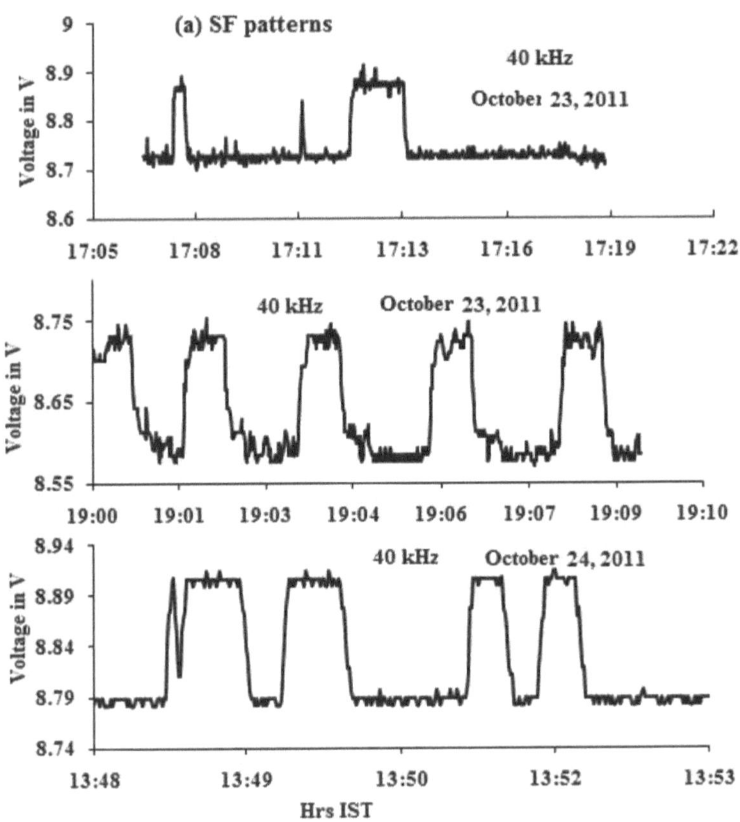

Figure 6.6(a) Typical records of sawtooth fadeout (SF) patterns

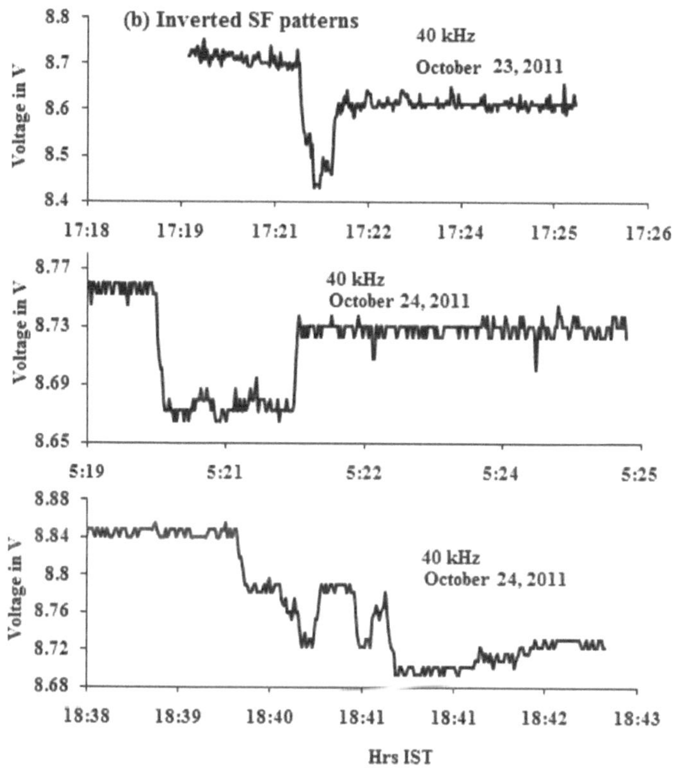

Figure 6.6(b) Typical records of inverted SF patterns at 40 kHz atmospherics

6.3 Practical uses of VLF Research

Lighting is the most devastating natural phenomena and it is destroying many things including lives per year. while the radio signature (like a fingerprint) of a lightning strike can be detected around the world, by receiving VLF waves, we can determine the exact location of most lightning strokes on the whole planet, with just a small number of VLF receivers scattered around [34-35].

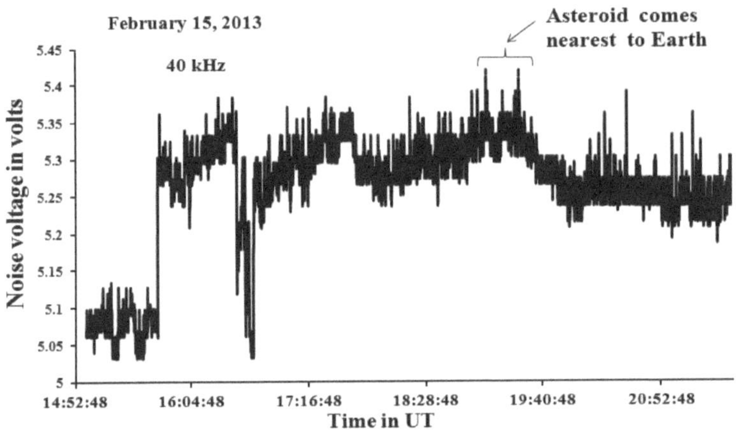

Figure 6.7 The variation of noise voltage at 40 kHz atmospherics with the arrival of the asteroid

6.3.1 Lightning Geo-location and **Lightning-Ionosphere Impacts**

Certain kinds of lightning can cause a brief discharge tens of miles above the cloud, which is called sprites. Actually, some lightning can even accelerate electrons to speeds enormously close the speed of light, producing gamma rays that we usually only see in association with collapsing stars and black holes. Lightning charge transfers of hundreds of Coulombs may be required for consistency with theories for sprite optical intensity and to account for the ELF Q-burst intensity [31]. Lightning causal to elves has a much flatter ("white") ELF spectra, and though associated with the very largest +CGs (often >150 kA), exhibits much smaller charge moments (< 300 C-km).

6.3.2 Global and Ground-to-Satellite Communications

VLF signal can be sent around the whole planet since VLF waves are guided by both the Earth and the ionosphere. VLF signals may be used for global communications but unfortunately to make VLF waves with an antenna, we need an antenna comparable to a wavelength, which would be very large. The Stanford VLF group is exploring an alternative method of VLF wave generation, which utilizes the same currents that make the northern lights, to turn the upper atmosphere into a big radiating antenna [32-33]. For Ground-to-satellite communications (or vice versa) VLF signal has to be used. Since the ionosphere is rapidly changing VLF waves give us the only consistent and continuous means of monitoring the lower ionosphere, and thereby, inferring communications outages [34].

6.3.3 Safety of Satellite Electronics

Many kinds of energetic charge particle are trapped by Earth magnetic field (Van Allen radiation belts). When satellites commence into space, they are immediately attacked by these energetic charge particles and after some time it's can damage the satellite electronics. Actually, these in periods of strong solar activity, the number of energetic particles can go through the roof, and stay there for years. VLF waves from lightning have been shown to be a key driver of the removal of these trapped particles, by propagating as a `whistler' wave through the magnetosphere [35].

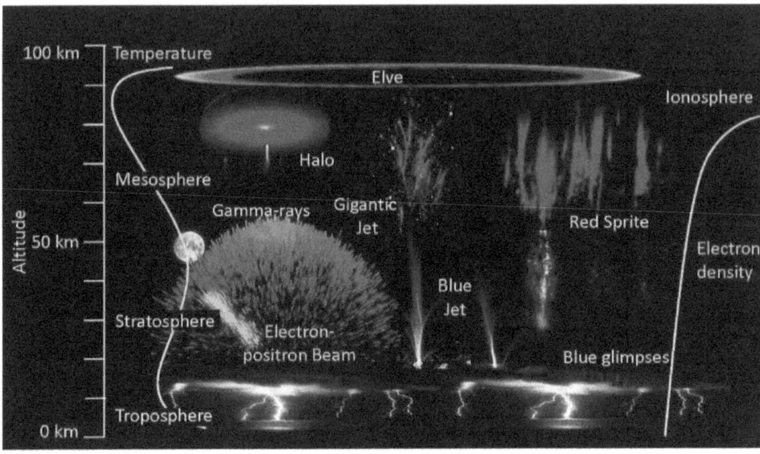

Figure 6.9 Illustration of upper atmospheric Transient Luminous Event phenomena of sprites, jets and elves [36]

6.4 Discussions

In association with the Indonesia earthquake followed by major cities of India, we have analyzed the effect on VLF atmospherics at 27 kHz. Figure 6.4 represents the characteristic variation in the atmospheric records at Kalyani University campus. The left-hand side record of atmospherics exhibits the typical variation in the noise level associated with a local nor'wester. The record shows a sudden enhancement of the noise level which is then repeated at higher Y-scale on April 10, 2012. Under this continued disturbed condition of the atmosphere on April 11, 2012 the violent earthquake was experienced at Indonesia and subsequently at different parts of India. Over Kolkata and its adjoining areas like Kalyani the shock was experienced at 14:14 IST which was later repeated once again with relatively weak strength. It appears that in the present case, the earthquake produces induced gravity wave perturbations [11, 12] in the ionospheric D-region during daytime producing the most irregular and asymmetric fluctuation in the noise level after the event [13, 14]. It is seemed that even when the violent earthquake occurred at a sufficiently remote location from Kalyani the severe disturbances may be responsible to influence our lower ionospheric layers and may even create problems in communication system temporarily.

The peculiar short period fading pattern in the reducing atmospherics level is indicative suggesting a change of electron density in the lower ionosphere of the earth [16] and thus affecting the ionospheric propagation. The signal received at 40 kHz further revels a rapid movement of the voltage levels with a sudden enhancement before the greatest transit of Venus again supports our view that the earth's ionosphere is affected in course of the contacts. Sawtooth fadeouts type of effects observed on the field strength of the received VHF signal and also at 40 kHz atmospherics level during the period of major Orionid meteor showers of October 2011 is presumed to be caused by over dense meteor trails in the propagation path involved. It may be interpreted by considering the altitude of formation of the ionized trail.

References

1. A. B. Bhattacharya and J. Pandit, *Ionosphere and its transient variations*, Lap Lambert Academic Publishing, Germany, 2014.

2. S. Joardar, S. Bhattacharyya, A. B. Bhattacharya and C. R. Datta, "Radio Astronomy and Supersynthesis: A Survey," *Progress In Electromagnetics Research B.*, 22, 73-102, 2010.

3. S. K. Sarkar, A. B. Bhattacharya and A. K. Sen, "On some characteristics of tropical atmospherics due to monsoon clouds", *Archives for meteorology, geophysics, and bioclimatology*, **31**, 411-419, 1982.

4. T. V. Prabha, A. Khain, R. S. Maheshkumar, G. Pandithurai, J. R. Kulkarni, M. Konwar and B. N. Goswami, "Microphysics of pre-monsoon and monsoon clouds as seen from in situ measurements during CAIPEEX". *Journal of the Atmospheric Sciences*, 68, 1882-1901, 2011.

5. M. W. Chiplonkar and R. N. Karekar, "The sunrise and sunset effect on the level of the atmospherics activity on the 27 kHz band", *Journal of Atmospheric and Solar Terrestrial Physics*, 25, 23-24. 1963.
6. C. Grandt, "Thunderstorm monitoring in South Africa and Europe by means of very low frequency sferics", *Journal of Geophysical Research*, 97, 18215-18226. 1992.

7. A. B. Bhattacharya, B. K. Dutta and R. Bhattacharya, "Some distinct effect of tropical monsoon clouds as derived from atmospherics", *Theoretical and Applied Climatology*, 50, 83-92, 1994.

8. A. B. Bhattacharya, "Multi-technique studies of nor'wester using meteorological and electrical parameters", *Annals of Geophysics*, **12**, 232 – 239, 1994.

9. T. G. Wood and U. S. Inan, "Long-range tracking of thunderstorms using sferic measurements", *Journal of Geophysical Research*, 107, 4553-4561, 2002.

10. M. W. Chiplonkar and G. G. Puranik, "Further analysis of VLF atmospherics data at Pune, India", *Indian Journal of Radio & Space Physics*, 5, 100-103, 1976.

11. F. F. Evison and D. A. Rhoades, "Demarcation and scaling of long-term seismogenesis", *Pure and Applied Geophysics*, 161, 21-45, 2004.

12. D. Shanker, R. B. S. Yadav and H. N. Singh, "On the seismic risk in the Hindukush-Pamir-Himalaya and vicinity", *Current Sciences*, 92, 1625-1630, 2007.

13. Mona Lisa, A. A. Khwaja and M. Q. Jan, "The 8 October 2005 Muzaffarabad earthquake: preliminary seismological

investigations and probabilistic estimation of peak ground accelerations", *Current Sciences*, 94, 1158–1166, 2008.

14. A. B. Bhattacharya, S. K. Kar, M. K. Chatterjee and R. Bhattacharya, "Long-period fading in atmospherics during severe meteorological activity and associated solar geophysical at low latitudes", *Annals of Geophysics*, 6, 183-188, 1998.

15. A. Chapman and J. Horrocks, "The Transit of Venus and the 'New Astronomy' in early seventeenth-century England", *Quarterly Journal of the Royal Astronomical Society*, 31, 333-357, 1990.

16. R. Bhattacharya, A. Bhoumick, A. Bhattacharya, S. Pal and A. B. Bhattacharya, "Changes in the surface UV radiation and sferics over Kalyani (West Bengal) during the solar eclipse on 22 July 2009", *Indian Journal of Radio & Space Physics*, 39, 132-137, 2010.

17. R. Woo and A. Ishimaru, "Eddy diffusion coefficient for the atmosphere of Venus from radio scintillation measurements", *Nature*, 289, 383-384, 1981.

18. C. T. Russell, T. L. Zhang, M. Delva, W. Magnes, R. J. Strangeway and H. Y. Wei, "Lightning on Venus inferred from whistler-mode waves in the ionosphere", *Nature*, 450, 661-662, 2007.

19. Q. Yin, S. B. Jacobsen and K. Yamashita, "A short time scale for terrestrial planet formation from Hf- W Chronometry of meteorites" *Nature*, 418, 949-951, 2002.

20. J. Rendtel and R. Arlt, *Editors, Handbook for Meteor Observations; Chapter 9, Analyses and Calculations*, International Meteor Organization, ISBN 978-2-87355-020-2, 2009.

21. B. Saha and A. K. Sen, "Long period fading of LF radio signals and geomagnetic activity during major meteor showers," *Annales de Geophysique.*, 36, 61-66, 1980.

22. S. P. Gupta, "Ionization layers over the magnetic equator during meteor shower days," *Advances in Space Research*, 10, 105–108, 1990.

23. B. Saha and A. K. Sen, "Attenuation of LF radio signals due to meteor trains," *Annales de Geophysique*,. **33**, 503-511, 1977.

24. A. K. Sen and B. Saha, "Radio wave propagation and ionization of meteoric constituents," *Nature*, 255, 313-316, 1975.

25. E. Asphaug, S. J. Ostro, R. S. Hudson, D. J. Scheeres and W. Benz, "Disruption of kilometre-sized asteroids by energetic collisions", *Nature,* 393, 437–440, 1998.

26. E. Asphaug, E. V. Ryan and M. T. Zuber, Asteroid interiors. In *Asteroids III* (W. F. Bottke Jr. et al., eds.), Univ. of Arizona, Tucson. 2002.

27. J. Bange, "An estimation of the mass of asteroid 20-Massalia derived from the Hipparcos minor planets data", *Astronomy and Astrophysics,* 340, L1–L4, 1998.

28. M. J. S. Belton, C. R. Chapman, P. C. Thomas, M. E. Davies, R. Greenberg, K. Klassen, D. Byrnes, L. D'Amario, S. Synnott, T. V. Johnson, A. McEwen, M. J. Merline, D. R. Davis, J. M. Petit, A. Storrs, J. Veverka and B. Zellner, "Bulk density of asteroid 243 Ida from the orbit of its satellite Dactyl", *Nature,* 374*,* 785–788, 1995.

29. W. F. Bottke, M. C. Nolan, R. A. Kolvoor and R. Greenberg, "Velocity distribution among colliding asteroids", *Icarus,* 107*,* 255–268, 1994.

30. W. F. Bottke, A. Morbidelli, R. Jedicke, J. M. Petit, H. Levison, P. Michel and T. S. Metcalfe, "Debiased orbital and absolute magnitude distributions of near-Earth objects", *Icarus*, 156, 399-433, 2002.

31. R. D. Hunsucker, J. K. Hargreaves, *The high-latitude ionosphere and its effects on radio propagation*, Cambridge University Press, 2002. ISBN 0-521-33083-1.

32. H. J. Christian, R. J. Blakeslee, D. J. Boccippio, W. L. Boeck, D. E. Buechler, K. T. Driscoll, S. J. Goodman, J. M. Hall, W. J. Koshak, D. M. Mach, M. F. Stewart, “Global frequency and distribution of lightning as observed from space by the optical transient detector”, *Journal of Geophysical Research*, 108, ACL 4-1–ACL 4-15, 2003.

33. D. R. Macgorman, D. W. Burgess, V. Mazur, W. D. Rust, W. L. Taylor, and B. C. Johnson, “Lightning rates relative to tornadic storm evolution on 22 may 1981”, *Journal of Atmospheric Sciences*, 46, 221-251, 1989.

34. R. A. Helliwell, *Whistlers and Related Ionospheric Phenomena*, Stanford University Press, Stanford, California, U.S.A, 1965.

35. U. Inan, N. Lehtinen, R. Moore, K. Hurley, S. Boggs, D. Smith, and G. Fishman, “Massive disturbance of the daytime lower ionosphere by the giant g-ray flare from magnetar SGR 1806-20”, *Geophysical Research Letters*, 34, L08,103, 2007.

36. E. R. Williams, “The positive charge reservoir for sprite-producing lightning”, *Journal of Atmospheric and Solar-Terrestrial Physics*, 60, 689-692, 1998.

Chapter 7

ESTIMATION OF THE FREQUENCY & COMPUTATIONS ON THE DRAKE EQUATION

7.1 Spontaneous Leakage of Artifacts

In recent years, interest has been developed among researchers, to search traces of extraterrestrial intelligences within the solar system. It has already been reported that, within the time of existence of our planet, about 10^4 stars capable of having inhabited planets, approached the Sun to distances within 1.5 pc (1 pc = 3.09 x 10^{13} km). It can be covered by space probes using present day technology [1 - 3]. Early works have attempted on a search for artifacts which are in orbit, on the Earth or on asteroids in a scattered way [4, 5]. It seems that this list should also include by elaborately considering orbits of different planets and their moons as well.

In this chapter, we have considered the orbital characteristics of different planets for estimating the frequency of the artifacts likely to fall down to earth and other planets. Space activities largely contribute to a long duration pollution of the interstellar medium [6]. In a similar way, the interplanetary space of other inhabited planetary systems could contain artifacts. The spontaneous leakage of artifacts into the interstellar medium in absence of interstellar flights is also reasonably expected.

This is mainly because of the following three reasons:

(1) A considerable portion of any large artifacts would be ejected by gravitational interaction with the planets. According to computer simulations of the asteroid and comet motion, 10-30 percent of small bodies leave the Solar system [7 - 9].

(2) Collisions between artifacts or their explosions (like spontaneous explosions of Earth satellites) in the outer parts of the planetary system could accelerate their debris up to hyperbolic velocities.

(3) Light pressure expels micron-sized debris particles (*e.g.* from rocket engines) out of the planetary system.

Hence, technical activities even within a planetary system led to a diffusion of artifacts into the interstellar medium. If there are alien artifacts between the stars, some of them are likely to fall down to Earth at some time. So, it is interesting to estimate the frequency of such events.

7.2 Theoretical Considerations

We have considered some mathematical equations so that we can calculate the probability of the fall on the Earth for artifacts.

The velocity (v) of all artifacts relative to the Sun in interstellar medium is given by,

$$v = \sqrt{\left[\frac{\{X^2 + Y^2 + Z^2 + (\sigma_1^2 + \sigma_2^2 + \sigma_3^2)\}}{3}\right]} \qquad (7.1)$$

where,

σ_1, σ_2, and σ_3 are the orthogonal dispersions in velocity of nearby stars;

X, Y, and Z are the components of the velocity vector of the Sun relatively to nearby stars [10].

Putting the values of,

σ_1= 38 km/s, σ_2= 24 km/s, σ_3 = 18 km/s

and X = 9 km/s, Y = 12 km/s, Z = 7 km/s,

we get, v = ***32.48 km/s***

Now, the effective radius of the Earth's orbit and the effective radius of Earth itself is given by,

$$A = a\sqrt{\left\{1 + \left(\frac{V}{v}\right)^2\right\}} \quad (7.2)$$

$$R_e \approx R\sqrt{\left\{1 + \left(\frac{u^2}{<v_a^2>}\right)\right\}} \quad (7.3)$$

respectively.

where,

a is the radius of the Earth orbit = 1.5x10^8 km (in average);

V is the escape velocity – 42.1 km/s at a = 1.5x10^8 km distance from the Sun.

R is the Earth's radius = 6371 km

u is the geocentric escape velocity of Earth = 11.2 km/s.

and $< v_a^2 >$ is the average square geocentric escape velocity of artifact which is given by,

$$< v_a^2 > = v^2 + 1.5\,V^2 \quad (7.4)$$

The probability of the fall of an artifact on the Earth at the distance of $r \leq a$ from the Sun is,

$$\omega = \left(\frac{R_e}{a}\right)^2 \quad (7.5)$$

Now, the number density (ρ) of artifact is given by,

$$\rho = \rho_0 \gamma e M C / m \qquad (7.6)$$

where:

ρ_0 = 4.43x10^{-42} km^{-3} is the stellar density near the Sun [11];

γ = 0.3 is the fraction of stars with planets among the nearby stars.

M = 2.3x10^{21} kg is the mass of potential raw material for the artifact manufacturing in the planetary system (the total mass of asteroids in the Solar System);

C = Part of raw material transformed into the interstellar artifacts; m is the typical mass of artifact; and

e = Fraction of planetary systems generating interstellar artifacts among the nearby planetary systems.

If v_h is the heliocentric velocity of nearby stars, then,

T is the age of the planet,

v^* is the artifacts injection velocity and

T_{max} is the maximum duration of exposure such that,

T = 1.43 x 10^{17} s and $v^* \sim$ 10 km/s

$$L = v^* \times T_{max} = 3 \times 10^{13}\ km \qquad (7.7)$$

Then the maximum number of stars which could infect the planet is,

$$N \sim \pi \rho_0 \gamma v_h T L^2 \qquad (7.8)$$

and the frequency of Extraterrestrial Intelligence artifact falls is,

$$f = \pi \rho r v \omega A^2 \approx 3.5 \times 10^{-11}\ eC/m \quad (7.9)$$

Taking m = 0.1, the upper limit of f could be calculated.

Now, e and C should be less than 1 i.e. $e \leq 1$ and $C \leq 1$.

So, putting the maximum value of e and C i.e. 1, we can estimate the frequency of artifact which is, $\boldsymbol{f = 3.5 \times 10^{-11}}$

If we put $m = 1$ kg, then, the average time between the falling of such ETI artifacts becomes, $\frac{1}{f} \geq 91$ yr.

Now Let, $e = 10^{-2}$, $C = 10^{-2}$ and $m = 0.1$ kg.

Then, the Earth could accumulate about five thousand of 0.1kg-artifacts during 4.5×10^9 years.

The probability of survival conditions of artifact is given by,

$$W = \sqrt{\frac{1 - \left(v_h^2 - v_e^2 - {}^{9.2}/_{\sigma}\right)}{2 v_h v_e}} \quad (7.10)$$

Where, v_h is the heliocentric velocity of artifact near the Earth and it is related by the equation,

$$v_h = \sqrt{(u^2 + v^2 + V^2)} \quad (7.11)$$

using the appropriate values of u, v and V, we get,

$\boldsymbol{v_h} = \mathbf{54.33}$ **km/s** and

the orbital speed of the Earth, $\boldsymbol{v_e} = \mathbf{29.78}$ **km/s**

7.3 Estimation of Possible Artifacts

In our estimation of the possible artifacts, we have considered the planets Mercury, Venus, Earth, Mars, Jupiter and Saturn. The orbital characteristics of all these planets are given in Table 7.1 and table 7.2. We estimate the probability of the artifacts fall on the Earth and the other five planets due to the Sun and draw the histogram shown in Figure 7.1. Figure 7.2 shows the percentage (%) contributions of falling artifacts.

Table 7.1 Comparison of orbital characteristics of Mercury, Venus and Earth [12]

Planet	**Mercury**	**Venus**	**Earth**
Aphelion in km	69×10^6	108×10^6	152×10^6
Perihelion in km	46×10^6	107×10^6	147×10^6
Semi-major axis in km	57×10^6	108×10^6	149×10^6
Radius of the orbit (a) in km	5.79×10^7	1.08×10^8	1.5×10^8
Mean radius in km	2,439.7	6,051.8	6,371.0
Mass in kg	3.30×10^{23}	4.87×10^{24}	5.97×10^{24}
Mean density in gm/cm^3	5.427	5.204	5.515
Geocentric Escape Velocity (u) in km/s	4.25	10.46	11.186

Escape Velocity at orbital distance (V) in km/s	67.88	49.70	42.1
V in km/s	32.48	32.48	32.48
$A = a\sqrt{\left\{1 + \left(\frac{V}{v}\right)^2\right\}}$	1.34×10^8	1.97×10^8	2.45×10^8
Effective Radius of the Planet R_e in km	2442.464	6120.956	6477.444
Average square of Geocentric Velocity $< v_a^2 >$	7966.492	4760.085	3713.565
$\omega = \left(\frac{R_e}{a}\right)^2$	1.78×10^9	3.21×10^9	1.86×10^9

Table 7.2 Comparison of orbital characteristics of Mars, Jupiter and Saturn [12]

Planet	**Mars**	**Jupiter**	**Saturn**
Aphelion in km	249×10^6	816×10^6	$1,513\times10^6$
Perihelion in km	206×10^6	740×10^6	$1,353\times 0^6$
Semi-major axis in km	227×10^6	778×10^6	$1,433\times 0^6$
Radius of the orbit (a) in km	2.27×10^8	7.78×10^8	1.43×10^9

Mean radius in km	3386.2	69,911	57,325
Mass in kg	6.42×10^{23}	1.90×10^{27}	5.68×10^{26}
Mean density in gm/cm^3	3.9335	1.326	0.687
Geocentric Escape Velocity (u) in km/s	5.027	59.5	35.5
Escape Velocity at orbital distance (V) in km/s	34.28	18.51	13.66
V in km/s	32.48	32.48	32.48
$A = a\sqrt{\left\{1+\left(\frac{V}{v}\right)^2\right\}}$	3.30×10^8	8.95×10^8	1.55×10^9
Effective Radius of the Planet R_e in km	3401.351	126160.7	79929.18
Average square of Geocentric Velocity $< v_a^2 >$	2817.628	1568.881	1334.844
$\omega = \left(\frac{R_e}{a}\right)^2$	22.45×10^9	2.63×10^8	3.12×10^9

From both the figures, it is clear that contribution of artifacts to Jupiter is maximum among all the planets considered. Geocentric escape velocity (u) of the six planets of the Sun as well as, the square root of the average square geocentric velocities $< v_a^2 >$ have been calculated and shown in Table 7.3.

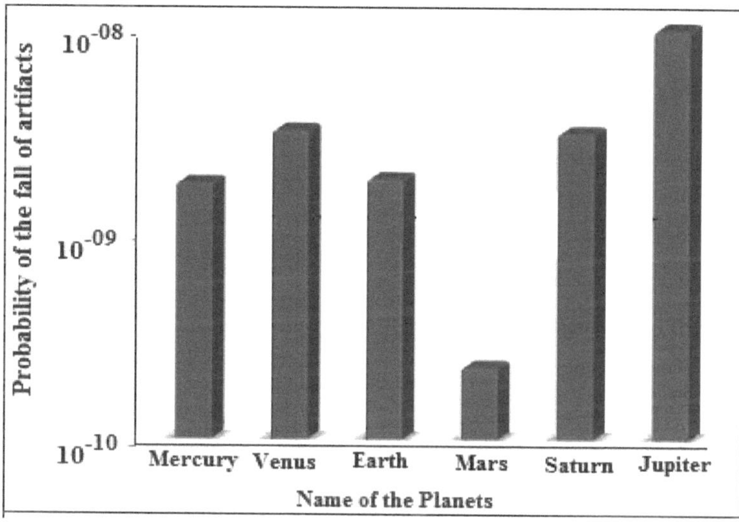

Figure 7.1 Probability of the fall of artifacts on Earth and other planets [13]

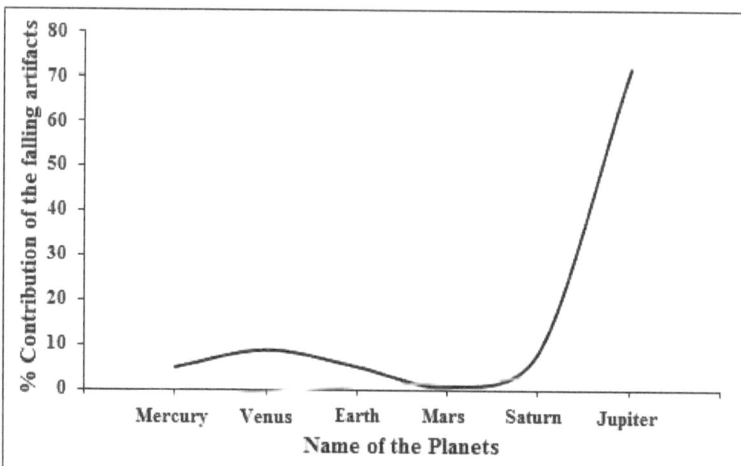

Figure 7.2 Percentage contributions of various Planets in the falling of artifacts [14]

Table 7.3 The Geocentric velocity (u) and the square root of average square geocentric velocity (v_a) for various planets including Earth [15]

Planets	u in km/s	u^2	v_a in km/s	$< v_a^2 >$
Mercury	4.25	18.0625	89.255	7966.492
Venus	10.46	109.4116	68.993	4760.0854
Earth	11.186	125.1266	60.939	3713.5654
Mars	5.026927491	25.27	53.081	2817.628
Jupiter	59.5	3540.25	39.609	1568.88055
Saturn	35.5	1260.25	36.535	1334.8438

Table 7.3 shows that the values of v_a are decreasing from the planet Mercury to the planet Saturn but it is not true for u. It is noted that the value of u is increasing from Mercury to the Earth. Then it diminishes for the planet Mars and again it increases for the other two planets considered. It appears from the table that, the value of u is maximum for Jupiter and minimum for Mercury while the value of v_a is maximum for Mercury and minimum for Jupiter. The variations of u and v_a corresponding to all the six planets have taken into account and are shown in Figure 7.3.

A plot of u^2 in logarithmic scale *vs.* $< v_a^2 >$ is made in Figure 7.4 which exhibits an interesting characteristic variation. The figure clearly shows that the variation of u^2 with $< v_a^2 >$ is non-linear in characteristics.

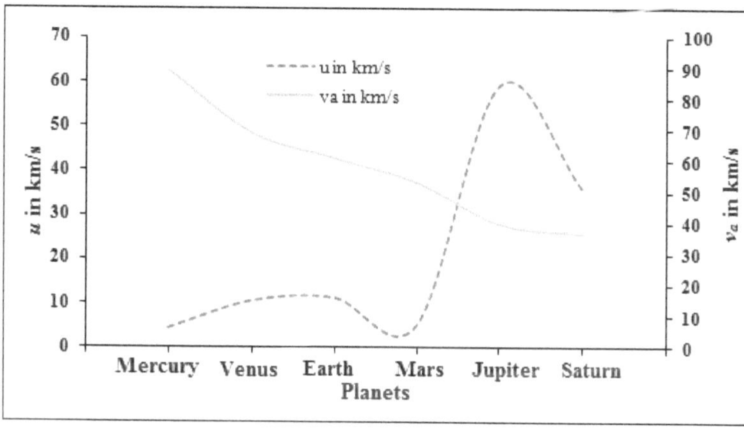

Figure 7.3 variations of u and v_a for various planets

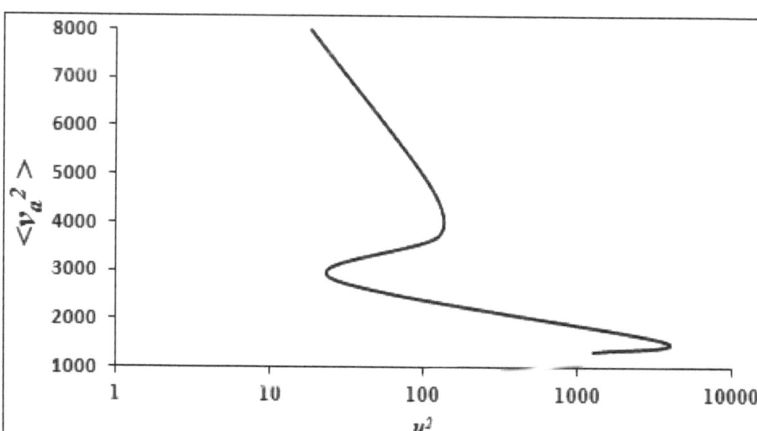

Figure 7.4 Plot of u^2 in log scale against $< v_a^2 >$ [16]

The effective radius R_e as derived from equation (7.3), can also be obtained from,

$$R_e = Rk \quad (7.11)$$

where,

$$k = \sqrt{1 + {u^2}/{< v_a^2 >}} \quad (7.12)$$

In Table 7.4 we have shown the calculated values of k as obtained by using above equation and also the corresponding planet's radius R.

Table 7.4 Radii of the planets and calculated values of k [17]

Planets	R in km	k
Mercury	2,439.70	1.001133013
Mars	3386.2	1.004474259
Venus	6,051.80	1.011427317
Earth	6,371.00	1.016707659
Saturn	57,325	1.394316304
Jupiter	69,911	1.804590059

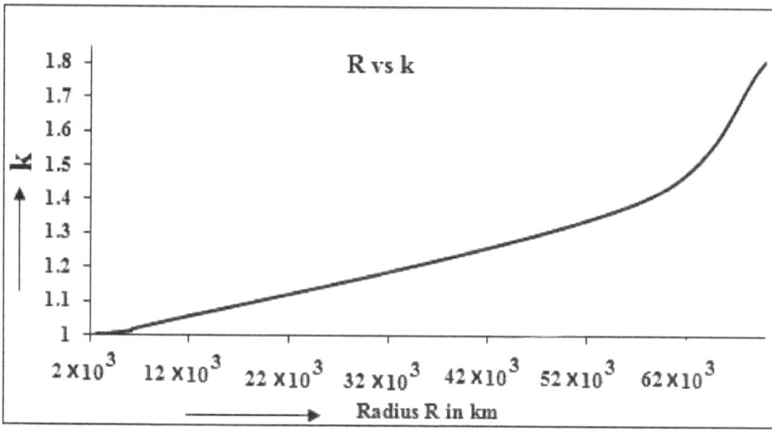

Figure 7.5 Plot showing the variation of R and k [17]

This variation prompted us to re-examine the radius of the orbit (a) with the corresponding effective radius of different planets (R_e).

A plot of a *vs* R_e is made in Figure 7.6(a) and their normalized values in Figure 7.6(b). From these figures we can compare the values of a and R_e of different planets with reference to Earth.

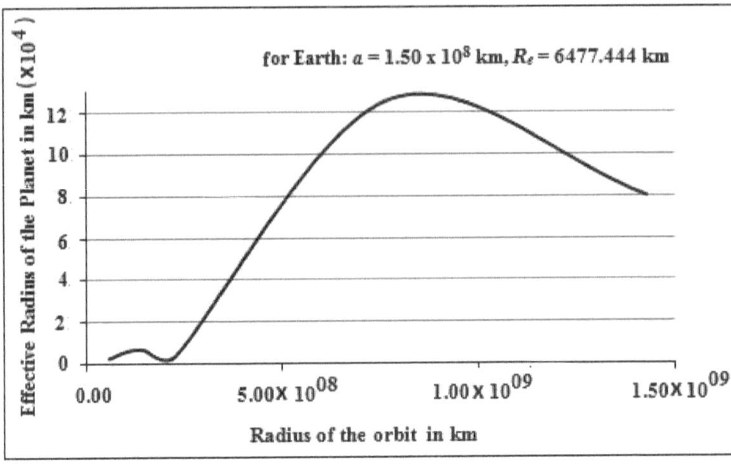

(a)

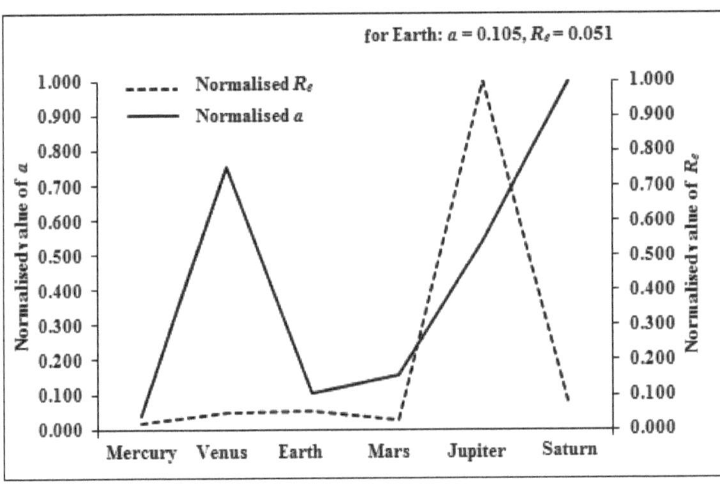

(b)

Figure 7.6 (a) Plot of radius of the orbit (a) *vs.* effective radius of the Planets (R_e), (b) Normalized values of a *vs* R_e

In Table 7.5, we have calculated the heliocentric velocity of nearby stars with the maximum number of stars which could infect the planets. The table shows that, the order of the maximum number of stars for all the planets are same (~ 10^4) but the heliocentric velocity varies from 50 to 75 km/s. This indicates that the heliocentric velocity (v_h) has not a major role in counting the maximum number of stars that could infect the planets.

Table 7.5 Heliocentric velocity of nearby stars and the corresponding maximum number to infect the planets [18]

Planets	**Heliocentric velocity of nearby stars** $v_h = \sqrt{(u^2 + v^2 + V^2)}\ km/s$	**Maximum number of stars which could infect the planet** $N \sim \pi \rho_0 \gamma v_h T L^2$
Mercury	75.37	4.0 x10^4
Venus	60.29	3.2 x10^4
Earth	54.34	2.9 x10^4
Mars	47.49	2.5 x10^4
Jupiter	70.27	3.7 x10^4

7.4 Theoretical considerations of the Drake equation

The Drake equation [19] is an attempt to establish the number of intelligent civilizations that existed in the Milky Way galaxy. The equation is composed of seven terms. Out of which the first six are used to compute the rate at which intelligent civilizations are being created and the final term identifies how long each last on an average as a broadcasting civilization [20-22]. It is, however, inapplicable to civilizations in other galaxies as they are too distant to be able to communicate through radio signals successfully. In this chapter we have first determined the two extreme values and thereby extracting the realistic parameters the probable number of broadcasting civilizations has been estimated. The limitations of the Drake equation are critically focused and its modifications have suggested.

According to Drake equation if the number of broadcasting civilization is denoted by N, then,

$$N = R \times n_e \times f_p \times f_l \times f_i \times f_c \times L \qquad (7.13)$$

where,

R = Average rate of formation of suitable stars (stars/year) in the Milky Way galaxy

n_e = Average number of habitable planets per star

f_p = Fraction of stars that form planets

f_l = Fraction of habitable planets (n_e) where life emerges

f_i = Fraction of habitable planets with life where intelligent may be evolved

f_c = Fraction of planets with intelligent life capable of interstellar communication

L = Years a civilization remains detectable.

Drake used the following values in expression (7.13) for identifying the number of civilizations:

R = 10, f_p = 0.5, f_i = 0.01, L = 10,000

n_e = 2.0, f_l = 1.0, f_c = 0.01,

When all those proposed values were substituted in the equation, the value of N comes out as 10; *i. e.* the number of broadcasting civilizations in our galaxy appears to be 10.

As a conservative estimate one may proceed with a belief that suitable planets are rare, life seldom becomes intelligent and intelligent civilizations do not last for a longer period. Under that restricted situation, the following typical values may get priority for computation:

R = 5/year, f_l = 0.13, L = 1,000 years

n_e = 0.01, f_i = 0.001,

f_p = 0.5, f_c = 0.01,

when we get by using Equation (7.13),

$$\boldsymbol{N = 0.0000325}$$

The chosen data set obviously concludes that we are almost alone in our galaxy. But for a more positive computed result, if we take,

R = 8/year,	f_l = 0.33,	L = 10,000 years
n_e = 2.0,	f_i = 0.01,	
f_p = 0.5,	f_c = 0.01,	

we get,

$$\boldsymbol{N = 2.64}$$

The results indicate that only two communicative civilizations exist in our galaxy at any given time, on an average.

Alternatively, if one is more optimistic and assumes that planets are common, life always arises when planets are favorable. 10% of civilizations become willing and able to communicate radio signals and then spread through their local star systems for 100,000 years. In fact, it is a very short period in geological time. We may go through the following computation:

R = 10/year,	f_l = 1.0,	L = 100,000 years
n_e = 2.0,	f_i = 0.1,	
f_p = 0.5,	f_c = 0.1,	

we get,

$$\boldsymbol{N = 10,000}$$

The optimistic assumption thus indicates that there are quite a good number of civilizations in the galaxy. All the above three results under varying situations prompted us to make a rigorous calculation wherein R, n_e and L values we have changed but $f_p,\ f_l,\ f_i,\ and\ f_c$ values are kept unaltered to find the appropriate values of $\boldsymbol{N}$ under changing conditions.

In a more realistic manner, we prefer to choose:

$f_p = 0.5$,

$f_l = 1.0$,

$f_i = 0.01$,

$f_c = 0.01$,

as fixed values in our computation when the Drake equation can be expressed in a simplified form as,

$$N = 0.00005\ R \times n_e \times L \qquad (7.14)$$

The computed results obtained by using Equation (7.14) are summarized in Table 7.6, which reveal, at a glance, the probable number of broadcasting civilizations with the length of time for which such civilizations release detectable signal in space. It appears from the table that if the years a civilization remains detectable are taken as 500 and 1000 for two sets of data, the maximum probable number of civilizations becomes 1.

On the other hand, if the years a civilization remains detectable be taken as 5,000 and 10,000 for another two sets of data the probable number of civilizations becomes always greater than 1 with a maximum to 10.

Table 7.6 Probable number of civilizations (*N*) for different average rate of formation of suitable stars (*R*), average number of habitable planets per star (n_e) and years a civilization (*L*) remains detectable

Set - I			
R	n_e	L	N
5	1	500	**0.125**
5	2	500	**0.25**
10	1	500	**0.25**
10	2	500	**0.5**
5	1	1000	**0.25**
5	2	1000	**0.5**
10	1	1000	**0.5**
10	2	1000	**1.0**

Set - I			
R	n_e	L	N
5	1	5000	**1.25**
5	2	5000	**2.5**
10	1	5000	**2.5**
10	2	5000	**5.0**
5	1	10000	**2.5**
5	2	10000	**5.0**
10	1	10000	**5.0**
10	2	10000	**10.0**

In Figure 7.7, we have combined our computed results of probable number of broadcasting civilizations considering year of civilization from 500 to 10,000 for different fixed realistic values of f_p, f_l, f_i, *and* f_c. This estimation suggests that the probable number of civilizations transmitting radio signals may lie between 1 and 10 in the Milky Way Galaxy. As an optimistic and conservative approach if the civilizations transmitting radio signals for a link with other civilizations are taken as 5, then also we can expect successful radio signal reception by applying suitable techniques and instruments.

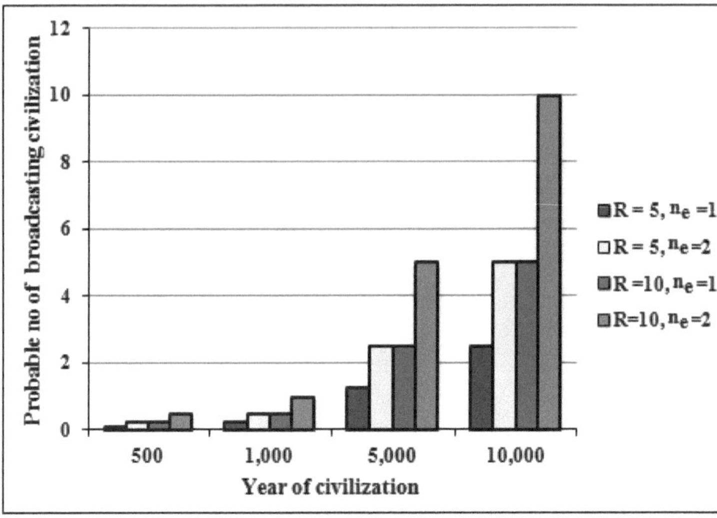

Figure 7.7 Probable occurrence of broadcasting civilizations with length of time

7.5 Discussion

The probability of the fall of the artifacts and their percentage contribution to different planets is a subject to be examined in greater details with the help of additional future information. Observational data supported by the future models developed can provide light towards falling of artifacts and eventually to the existence of extraterrestrial intelligence at our planetary systems. It has been reported that Mercury and Mars are less than 1 percent while that for the Venus and Saturn is greater than 1 [23]. There is a higher chance that the artifacts on the Earth may arise from Venus or Saturn than Mercury or Mars.

Some recent theories suggest that Venus and the Earth may have started out alike [24]. There might have been a lot of water on Venus and there might have been a lot of carbon dioxide on Earth. But all that was to change. On Earth, life in the oceans took in carbon dioxide and turned it into limestone. On Venus, 30% closer to the Sun, any oceans boiled away and the water vapor added to the runaway greenhouse effect. It is largely believed that microbes could survive and reproduce; floating in the thick, cloudy atmosphere and also widely protected by a sunscreen of sulfur compounds [25]. "Panspermia" the theory that life on Earth came from space, comets, or other planets, has an ancient history [26]. There are more promising non-classical SETI possibilities than the conventional search for radio/laser signals. But these approaches conflict with the mental habits of astronomers, geologists and geochemists in studying natural formations and processes. This habit factor leads most specialists to a priori rejection of search for alien artifacts on the surfaces of the planets and their moons. Invaluable information about the astro-dynamical aspect of paleohistology, particularly regarding the formation and extent of the artifacts as well as migration of small bodies in the solar planetary systems to be considered at a greater length for evaluating ETI problems.

The Drake equation is still not a very successful one due to some of the elements in the equation which would take too long to get an accurate sample for providing a scientifically sound explanation. It further appears that the equation should consider other factors of practical importance. As for example, the equation should be multiplied by how many times an intelligent civilization may occur on planets where it has once happened [27]. Even if an intelligent civilization comes to an end of its lifetime, life may still prevail on the planet for billions of years wherefrom the next civilization may be evolved. Hence several civilizations may appear and disappear during the life span of a single planet. If n_r be the average number of times a new civilization may reappear on the same planet where an early civilization has once appeared and ended, then the total number of civilizations in that planet would be $(1+ n_r)$, which is the actual reappearance factor and should be included in the Drake equation. However, it should be mentioned that the factor n_r depends on the cause of civilization extinction. In case of temporary uninhabitability like a nuclear winter, n_r should be relatively high but for permanent uninhabitability like stellar evolution, n_r may be almost zero. In fact, the Drake equation has many uncertainties in its parameters. Also identifying and categorizing the relevant parameters is the most fundamental task for the proper utilization of the equation and thereby to exploit the search for extraterrestrial intelligence and the likelihood of its existence. It should be pointed out that radio propagation in GHz frequency is largely by attenuation due to precipitation and so a successful reception of any signal in microwave band there is a need of a strong database [28].

Data Source

1. http://www.astrodigital.org/astronomy/drake_equation.html

References

1. A. B. Bhattacharya, S. Das, A. Sarkar, M. Debnath and A. Bhoumick, "A new technique for investigating extrasolar planets", *International Journal of Physics,* 4, 31-42, 2011.

2. J. Heidmann, "SETI - Programmes All Over the World (and further out)", *Journal of the British Interplanetary Society*, 48, 447-452, 1995.

3. A. B. Bhattacharya, D. K. Tripathi, A. Sarkar, A. Bhoumick, "Implementation of Radio Technique in Interstellar Radiation Field for Locating Distant Stars", *International Journal of Applied Engineering Research*, 2, 63-69, 2011.

4. V. Arkhipov, "Astro-dynamical aspect of paleohistology", *Research Institute on Anomalous Phenomena Bulletin*, 1(1), 7-10, 1994.

5. A. R. Martin, "Project Daedalus: bombardment by interstellar material and its effects on the vehicle", *Journal of the British Interplanetary Society, supplement*, S116-S121, 1978.

6. H. Heusmann, "Space Debris", *Spaceflight*, 35, 182-184, 1993.
7. M. Duncan, T. Quinn and S. Tremaine, "The formation and extent of the solar system cloud", *Astronomical Journal*, 94, 1330-1338, 1987.

8. P. Farinella, Ch. Froeschlé, C.Froeschlé, R. Gonczi, G. Hahn, A.Morbidelli and G.B. Valsecchi, "Asteroids failing into the Sun", *Nature*, 371, 314-317, 1994.

9. S. I. Ipatov, "Migration of Small Bodies to the Earth", *Astron. Vestn.*, 29, 304-330, 1995.

10. W. Allen, *Astrophysical Quantities*, The Athlone Press, London, 1973.

11. A. Tough, "A critical examination of factors that might encourage secrecy", *Acta Astronautica*, 21 (2), 97-102, 1990.

12. M. Mark, "Why SETI is science and UFOlogy is not", *Skeptical Inquirer*, 2004.

13. M. Chown, "The Alien Spotters". *New Scientist*, 28, 1997.

14. M. Redfern, "*Venus Clouds Might Harbor Life", BBC News*,

15. P. Curd, *Anaxagoras of Clazomenae: Fragments and Testimonia,* University of Toronto Press, Toronto, 2007.

16. F. Hoyle and C. Wickramasinghe, *Our place in the Cosmos, J. M. Dent, London*, 1993.

17. F. Hoyle and C. Wickramasinghe, *Astronomical origins of life, steps towards Panspermia,* Klewer Academic Publishers, 2000.

18. S. Arrhenius, *Worlds in the Making*, *Harper & Brothers, New York*, 1908.

19. F. Crick, *Life Itself. Its Origin and Nature*, *Simon & Schuster, New York,* 1981.

20. A. B. Bhattacharya, S. S. Banerjee and R. Bhattacharya, "A Roadmap for detecting Extraterrestrial Intelligent Life", *PIERS Proceedings*, Beijing, China, March 23-27, 1400-1403, 2009.

21. A. B. Bhattacharya, S. Das, A. Sarkar, M. Debnath and A. Bhoumick, "A new Technique for Investigating Extrasolar Planets", *International Journal of Physics*, 4, 31- 42, 2011.

22. A. B. Bhattacharya, S. Joardar and R. Bhattacharya, *Astronomy and Astrophysics*, Jones and Bartlett Publisher, Canada, 2010.

23. D. Charbonneau, L. Allen, S. Megeath, G. Torres, R. Alonso, T. Brown, R. Gilliland, D. Latham, G. Mandushev, F. O'Donovan and A. Sozzetti, "Detection of Thermal Emission from an Extrasolar Planet", *The Astrophysical Journal*, 626, 523-529, 2005.

24. K. L. Luhman, L. Adame, P. D'Alessio and N. Calvet, "Discovery of a Planetary-Mass Brown Dwarf with a Circumstellar Disk", *The Astrophysical Journal*, 635, 93-96, 2005.

25. J. Elliott, "A Post-Detection Decipherment Matrix: Acta Astronautica", *Journal of the International Academy of Astronautics*, AA, 2853, 2007.

26. A. E. Reines, and G. W. Marcy, "Optical Search for Extraterrestrial Intelligence: A Spectroscopic Search for Laser Emission from nearby Stars", *Publications of the Astronomical Society of the Pacific*, 114, 416-426, 2002.

27. S. K. Sarkar and A. Kumar, "Cloud attenuation and cloud noise temperature over some Indian eastern stations for satellite communication", *Indian Journal of Radio & Space Physics*, 36, 375-378, 2007.

28. A. B. Bhattacharya, D. K. Tripathi, A. Sarkar and A. Bhoumick, "Implementation of Radio Technique in Interstellar Radiation Field for Locating Distant Stars" *International Journal of Applied Engineering Research*, Dindigul, 2, 63-69, 2011.

Chapter 8

SUMMARY OF THE BOOK & SCOPE FOR FURTHER INVESTIGATIONS

8.1 Summary of this Book

Important results obtained from the present investigation done in this book have been summarized below:

Chapter 3: Investigation of Interstellar Signal using H-line Spectrometer

(i) Within the so-called Microwave Window, photons (the substance of electromagnetic communication) travel relatively unimpeded through the interstellar medium, at the speed of light. This is, in fact, the fastest possible speed making photons the fastest spaceships known to man. Thus, the Microwave Window, where natural noise is at a minimum, is a favored region for conducting radio astronomy research, including SETI. Low frequencies are obstructed by galactic noise, primarily due to synchrotron radiation while high frequencies are subject to quantum-effect emissions and the whole continuum experiences a 3 Kelvin background radiation level from the residual radiation of the Big Bang. These natural radiation sources limit our ability for detecting artificial emissions. Moreover, the Earth's own ocean of air generates spectral absorption and emission lines to draw a further curtain across our sky. However, there are a few relatively clear windows on the cosmos. Our eyes evolved to operate in two such windows. These windows will allow us to observe the stars and planets hopefully, if the proposed experiment can be performed with utmost care.

(ii) We have considered quite a good number of reported Star/Constellation (Kraus, 1986) and their corresponding distances for the present investigation as a first step. Using the values of the distances of the stars given in light year (Ly), we have calculated the time taken by the radio signal to reach the concerned Star and also the total time taken by the signal to back to the Earth again.

Figure 3.5 reveals the occurrences of selected stars at different distances while Figure 3.6 presents a scatter plot of the different time zones required by the radio signal to arrive at the earth as echo signal. It appears from Figure 3.5 that more than 60% of stars are situated up to a distance of 500 Ly. However, when Figure 3.6 is taken into analysis no prominent conclusion can be drawn except that the required time of arrival of the echo signal within 10 years period from its time of transmission is not sufficient in number. This prompted us to re-examine the response probability of the stars by transmitting radio signals through the two water holes using different transmitting powers. In Figure 3.7 we have plotted the probable number of stars to be detected by radio signal when transmitted power of the signal through the two 'water holes' are varied over a wide range. It appears from the figure that the number of stars corresponding to the two wavelengths 0.18 m and 0.21 m are almost overlapped and so for the selection of transmitting a signal corresponding to this water hole a mean of these two wavelengths may be chosen as a compromise.

In the above calculation, we find that depending on the transmitted power and hence on the radius (in Ly) the numbers of stars vary. It is interesting to note that when the radius is significantly small, the number of stars may even be less than one (as shown by * marks) in the Table 3.6. This indicates that a minimum value of the transmitted power is essential for detecting and identifying any star.

Considering the sky noise windows of the atmosphere we are suggesting the radio communication approach here. A sky noise diagram, indicating sky noise temperature against frequency, is shown in Figure 3.3. If the hydrogen line and the water line are considered, we find that the hydrogen (H) line is at 21 cm whereas the hydroxyl radical (OH) line is at 18 cm. When hydrogen is combined with hydroxyl radical, it produces water (H_2O) whose wavelength lies between 18 cm and 21 cm. The 21cm line is formed by an exciting quantum peculiarity in the nature of the electrons within the hydrogen atom.

Chapter 4: Observation of Galactic Radio Signal using H-Line at 1420 MHz

When we were receiving data of spectrum type round-the-clock and also noted many interesting characteristics here also. In Figure 4.5 (a) we plot a typical variation of the signal received by 1420 MHz hydrogen line Spectra-Cyber receiver (it is a scatter plot that received by the receiver). Here operating mode of the receiver in spectrum and the horizontal axis representing the frequency from 1419 MHz to 1421 MHz. it is interesting to see from the figure that the maximum voltage received at 1420.2 MHz. We extract the data from the save file and again plot this by connecting the dots points, we get another file which shown in the Figure 4.5 (b). Some other data sheet is also present here by Figure 4.6 and Figure 4.7. The corresponding dates the data also indicated in the figurers. It very interesting to note that in all type of data sheet, the maximum pick voltages happened on 1420.2 MHz frequency.

We were receiving data of continuum type round-the-clock and noted many interesting characteristics. In analyzing the received continuum data, we have classified them into five categories, *viz.* A, B, C, D and E, to some extent arbitrarily, depending upon their various physical shapes obtained. A typical continuum record at 1420 MHz observed in our laboratory at 17:03:52 UT on May 23, 2013 is exhibited in Figure 4.1. We extract data and then changes the time from UT to IST by adding 5:30 Hours in time scale. We again plot this using Excel file, which is shown in Figure 4.1a. Here we noticeably say that the maximum pick arises about 22:35:40 IST. The variations of amplitude occurred from 1.56 volts to almost 1.585 volts. By clicking we can set the background of the main window which drops a .bmp file for including In publication documents or an animated gif of the data. An X-axis feature permits km/sec for continuum record. Another feature permits to use either dots or connected lines for the plotting. Here X- axis represent frequency scale. The final feature is the Plot Capturing to allow whether or not a bmp graph file is created on each save of the scan. Data stored on disk can be played back in a "movie". Each screen of data is pulled from the disk and rapidly drawn on the Screen. The previous screen is then erased and the next file in sequence is displayed. The files to play are previously stored on the hard disk. The software has the ability to start data collection at a point and this can be chosen at convenient time. When completed with filling out the settings values the spectrometer moves to Online if the communications channel is open. The final option is the Input Frequency Shift which brings up a dialog box to know how much we want to shift the X- axis values. In Table 4.1, we have presented summary information of the Spectra-Cyber Software.

Chapter 5: Studying Transient Features through the Solar Corona by 406.7 MHz Radio Telescope

Observations of the solar corona in front of the solar disk have established a direct relationship between coronal features and solar wind. Regions of solar atmospheres known as coronal holes closely correspond to the high-speed component of the solar wind. The solar wind is able to generate significant heating of the lower thermosphere at high latitudes by direct particle precipitation and also by generating upper atmospheric ionization that may influence the global electric circuit. Due to coupling with the magnetosphere the solar wind drives ionospheric currents at high latitudes which in turn accelerate the neutral atmosphere. Some investigations have suggested a strong relationship between the electric field of the solar wind and a pressure phenomenon in the north Atlantic called as the North Atlantic Oscillation (NAO).

The corona is basically a part of the solar flare (Coronal mass ejections are often related with other forms of solar activity, most notably solar flares, although a causal relationship has not been established). The temperature of the corona is so high that it may throw some charged particles as solar wind. These may build up the electric field that again accelerates the charged particles. The relation of solar wind and coronal feature can be established by observing the solar corona in front of solar disk. Coronal holes closely communicate with high-speed component of solar wind and these solar wind particles can interact with upper atmosphere and may influence the global electric circuit at high latitude. The signal strengths of the coronal effect in our observed data were very high compared to the surface effect of the Sun. In fact, the signal level of surface effect of the Sun at the peak activity appears to vary in amplitude from 0.06 volts to 0.1 volt while those related to solar corona exhibit a change in between 7.0 to 7.2 volts. The solar wind also couples with magnetosphere and hence derives the ionosphere so that it again interacts with neutral atmosphere. These solar wind particles, i. e. the so-called charged particles interact with Earth's magnetic field and influence the geomagnetic indices. Due to the active role of the charged particles, the geomagnetic indices are enhanced at the time of solar bursts when the DST index reduces rapidly.

Chapter 6: Investigation on VLF Radio Signals and some Interrelated Parameters

In association with the Indonesia earthquake followed by major cities of India, we have analyzed effect on VLF atmospherics at 27 kHz. Figure 6.3 represents the characteristic variation in the atmospheric records at Kalyani University campus. The left-hand side record of atmospherics exhibits the typical variation in the noise level associated with a local nor'wester. The record shows a sudden enhancement of the noise level which is then repeated at a higher Y-scale on April 10, 2012. Under this continued disturbed condition of the atmosphere on April 11, 2012 the violent earthquake was experienced at Indonesia and subsequently at different parts of India. Over Kolkata and its adjoining areas like Kalyani the shock was experienced at 1414 IST which was later repeated once again with relatively weak strength.

Chapter 7: Estimation of the Frequency and Computations on the Drake Equation

The probability of the fall of the artifacts and their percentage contribution to different planets is a subject to be examined in greater details with the help of additional future information. Observational data supported by the future models developed can provide light towards falling of artifacts and eventually to the existence of extraterrestrial intelligence at our planetary systems. It has been reported that Mercury and Mars are less than 1 percent while that for the Venus and Saturn is greater than 1. There is a higher chance that the artifacts on the Earth may arise from Venus or Saturn than Mercury or Mars. Some recent theories suggest that Venus and the Earth may have started out alike. There might have been a lot of water on Venus and there might have been a lot of carbon dioxide on Earth. But all that was to change. On Earth, life in the oceans took in carbon dioxide and turned it into limestone. On Venus, 30% closer to the Sun, any oceans boiled away and the water vapor added to the runaway greenhouse effect. It is largely believed that microbes could survive and reproduce; floating in the thick, cloudy atmosphere and also widely protected by a sunscreen of sulfur compounds. "Panspermia" the theory that life on Earth came from space, comets, or other planets, has an ancient history. There are more promising non-classical SETI possibilities than the conventional search for radio/laser signals. But these approaches conflict with the mental habits of astronomers, geologists and geochemists in studying natural formations and processes. This habit factor leads most specialists to a priori rejection of search for alien artifacts on the surfaces of the planets and their moons. Invaluable information about the astro-dynamical aspect of paleohistology, particularly regarding the formation and extent of the artifacts as well as migration of small bodies in the solar planetary systems to be considered at a greater length for evaluating ETI problems.

The Drake equation is still not a very successful one due to some of the elements in the equation which would take too long to get an accurate sample for providing a scientifically sound explanation. It further appears that the equation should consider other factors of practical importance. As for example, the equation should be multiplied by how many times an intelligent civilization may occur on planets where it has once happened. Even if an intelligent civilization comes to an end of its lifetime, life may still prevail on the planet for billions of years wherefrom the next civilization may be evolved. Hence several civilizations may appear and disappear during the life span of a single planet. If n_r be the average number of times a new civilization may reappear on the same planet where an early civilization has once appeared and ended, then the total number of civilizations in that planet would be $(1 + n_r)$, which is the actual reappearance factor and should be included in the Drake equation. However, it should be mentioned that the factor n_r depends on the cause of civilization extinction. In case of temporary uninhabitability like a nuclear winter, n_r should be relatively high but for permanent uninhabitability like stellar evolution, n_r. may be almost zero. In fact, the Drake equation has many uncertainties in its parameters. Also identifying and categorizing the relevant parameters is the most fundamental task for the proper utilization of the equation and thereby to exploit the search for extraterrestrial intelligence and the likelihood of its existence. It should be pointed out that radio propagation in GHz frequency is largely by attenuation due to precipitation and so a successful reception of any signal in microwave band there is a need of a strong database.

8.2 Scope for Further Investigation

Within the so-called Microwave Window, photons (the substance of electromagnetic communication) travel relatively unimpeded through the interstellar medium, at the speed of light. This is, in fact, the fastest possible speed making photons the fastest spaceships known to man. Thus, the Microwave Window, where natural noise is at a minimum, is a favored region for conducting radio astronomy research, including SETI. Low frequencies are obstructed by galactic noise, primarily due to synchrotron radiation while high frequencies are subject to quantum-effect emissions and the whole continuum experiences a 3 Kelvin background radiation level from the residual radiation of the Big Bang. These natural radiation sources limit our ability for detecting artificial emissions. Moreover, the Earth's own ocean of air generates spectral absorption and emission lines to draw a further curtain across our sky. However, there are a few relatively clear windows on the cosmos. Our eyes evolved to operate in two such windows. These windows will allow us to observe the stars and planets hopefully, if the proposed experiment can be performed with utmost care.

The interstellar medium is basically the dust and gas that pervade interstellar space representing the matter that exists between the stars within a galaxy. The energy, in the form of electromagnetic radiation that occupies the same volume is known as the interstellar radiation field. The interstellar medium consists of a dilute mixture of ions, atoms, molecules, larger dust grains, cosmic rays, and also the galactic magnetic fields. It has been well reported that the matter consists of about 99% gas and 1% dust by mass. As a result of primordial nucleosynthesis, the gas is approximately 90% hydrogen and 10% helium by number of nuclei, with an additional heavier elements present in trace amounts, the stars of our Galaxy and presumably the stars in all other billions of galaxies in the Universe are born in interstellar clouds. It is mainly because of that search-strategy planning becomes a very complex one due to so much uncertainty arising for handling various parameters at the time of radio signal transmission through the interstellar medium. This may be minimized by repeating the experiment and giving due importance to the individual parameters taken into account.

The Drake equation is still not a very successful one due to some of the elements in the equation which would take too long to get an accurate sample for providing a scientifically sound explanation. It further appears that the equation should consider other factors of practical importance. As for example, the equation should be multiplied by how many times an intelligent civilization may occur on planets where it has once happened. Even if an intelligent civilization comes to an end of its lifetime, life may still prevail on the planet for billions of years wherefrom the next civilization may be evolved. Hence several civilizations may appear and disappear during the life span of a single planet. If n_r be the average number of times a new civilization may reappear on the same planet where an early civilization has once appeared and ended, then the total number of civilizations in that planet would be $(1 + n_r)$, which is the actual reappearance factor and should be included in the Drake equation. However, it should be mentioned that the factor n_r depends on the cause of civilization extinction. In case of temporary non-habitability like a nuclear winter, n_r should be relatively high but for permanent non-habitability like stellar evolution, n_r may be almost zero. In fact, the Drake equation has many uncertainties in its parameters. Also identifying and categorizing the relevant parameters is the most fundamental task for the proper utilization of the equation and thereby to exploit the search for extraterrestrial intelligence and the likelihood of its existence. It should be pointed out that radio propagation in GHz frequency is largely by attenuation due to precipitation and so a successful reception of any signal in microwave band there is a need of a strong database. After analyzing our collected data, we have planned to work in this line.

The study of the fine structure of the solar radio emission is a key to understand plasma processes in the solar corona. With a view to simplify the analyzing technique, we need to fine-tune the algorithm. This can help particularly when a large number of data is required to handle more systematically. In fact, a more appropriate solar burst identification strategy may be highly profitable at this time when the solar activity is picking up. Efforts are currently underway to calculate the response matrix for flare time intervals. In this regard, detailed spectral analysis and cross-calibration with other solar instruments are essential. We are planning to install on our server the response matrix software and generate response files for all flare time intervals. It is the same date but with different time in IST, taken into consideration by us for illustrating our own data as a typical case and thus have used as a conclusive comparison.

A solar burst is a huge explosion in the Sun's atmosphere which affects different layers, particularly photosphere, chromospheres, and corona of the sun. It is responsible for heating plasma to very high temperature (~10^7 K) and accelerating nucleons and heavier ions to near the speed of light. As a matter of fact, they generate radiation across the electromagnetic spectrum from radio waves to gamma rays. Most of the bursts occur in active regions around sunspots, where strong magnetic fields penetrate the photosphere to link the corona to the solar interior. Solar flares are powered by the sudden release of magnetic energy stored in the corona. A strong flare can cause coronal mass ejections (CME). X-rays and UV radiation emitted by solar flares can affect the ionosphere of the Earth when long-range radio communications may be disrupted. The occurrence of solar flares becomes more when the sun is active and less when the sun is quiet. Large flares are less frequent in comparison to smaller ones. In course of our observations all these features will be critically examined and will be reported in due course. The cosmic rays play a very vital role in the electrical properties of the atmosphere and the global electric circuit by manipulating atmospheric conductivity, ionospheric potential, vertical current and vertical potential gradients related to ionosphere. From a combination of available observing data and theoretical concept daytime and nighttime Martian electron density profiles when considered with the terrestrial electron density profiles, it is noted that in spite of many similarities between the terrestrial and Martian electron density profiles the main dissimilarity is the presence of a distinct ionization layer at a different height on Mars in comparison to the earth's ionospheric height which is produced by incoming cosmic rays. There is a pressing need for direct measurements of planetary atmospheric electrification. It is particularly true for Mars, to assess the risk for future manned and unmanned missions. Microphysical modeling found that lightning was most likely to occur in the ammonium hydro-sulphide clouds, but the investigation was restricted due to lack of experimental data on the electrical properties of ammonium hydro-sulphide and the related charge transfer. More ground-based laboratory observations and systematic study could improve our understanding of the charging mechanisms involved.

In spite of the existence of aerosol, ions, polar molecules and convection, it has become difficult to apply the terrestrial model of a global circuit to outer planets of the solar system mainly because of the probable absence of a conducting surface. A different electrical model for the gas giant planetary atmospheres may be attempted but further theoretical work is necessary to enhance our understanding first. The same or at least similar atmospheric electrical processes may act across the solar system but a clear concept is initially essential to implement it successfully. Lightning has been detected on many other planets but without their elaborate knowledge of electrical and physical properties we are only able to approach so far as quasi-terrestrial global circuits. Ion-mediated nucleation has been noted on Earth but no successful attempt has yet been made to identify it in other planetary atmospheres. A comparative approach with all the available future information may provide new insight for successful development of electrical global circuits for other planets where the earth's electrical global circuit may be taken as a ready reference.

The probability of the fall of the artifacts and their percentage contribution to different planets is a subject to be examined in greater details with the help of additional future information. Observational data supported by the future models developed can provide light towards falling of artifacts and eventually to the existence of extraterrestrial intelligence at our planetary systems. It has been reported that Mercury and Mars are less than 1 percent while that for the Venus and Saturn is greater than 1. There is a higher chance that the artifacts on the Earth may arise from Venus or Saturn than Mercury or Mars. Some recent theories suggest that Venus and the Earth may have started out alike. There might have been a lot of water on Venus and there might have been a lot of carbon dioxide on Earth. But all that was to change. On Earth, life in the oceans took in carbon dioxide and turned it into limestone. On Venus, 30% closer to the Sun, any oceans boiled away and the water vapor added to the runaway greenhouse effect. It is largely believed that microbes could survive and reproduce; floating in the thick, cloudy atmosphere and also widely protected by a sunscreen of sulfur compounds. "Panspermia" the theory that life on Earth came from space, comets, or other planets, has an ancient history. There are more promising non-classical SETI possibilities than the conventional search for radio/laser signals. But these approaches conflict with the mental habits of astronomers, geologists and geochemists in studying natural formations and processes. This habit factor leads most specialists to a priori rejection of search for alien artifacts on the surfaces of the planets and their moons. Invaluable information about the astro-dynamical aspect of paleohistology, particularly regarding the formation and extent of the artifacts as well as migration of small bodies in the solar planetary systems to be considered at a greater length for evaluating ETI problems.

We have noted that the effect of magnetic activity related to A_p and C_p values attains its peak level around the date of the peak activity of the Orionids shower. There is no such increase in the magnetic activity during the non-shower days. It has further been noticed that plot of K_p values and *AE* indices has interesting anomaly exhibiting an enhanced magnetic activity for K_p, like A_p and C_p, but a clear decrease of the magnetic activity of AE indices, even showing lowest during peak Orionid meteor showers. When we compare on the same platform the characteristic variations of all the four magnetic parameters (viz. A_p, C_p, K_p and *AE* indices) we note a significant and very prominent changes during shower days with non-shower time.

Our analysis indicates that the major meteor showers related to Orionids may be strong sources of geomagnetic disturbances. It was reported earlier that the auroral zone during geomagnetic storms of solar origin is subsequently redistributed to lower latitudes due to gravity wave perturbations. Present analysis reveals that the Orionid showers induce geomagnetic disturbances presumably due to dumping of the ionized debris of meteors which eventually give rise to gravity waves during their propagation from polar to equatorial region and thereby to play an important role for redistributing the ionization in the appropriate height of the ionosphere which occurs at the path midpoint of the received signal. In fact, the greater number of occurrences of SF effects at night may be due to the reflection of signal at nighttime from a higher height (~90 km) and the probability of formation of an ionized meteor trail at such height with particle line density is sufficient to affect the 40 kHz VLF propagation. The photo-detachment process may be operated in daytime to diminish the attachment coefficient, β_c. The higher value of β_c makes the duration of the observed SF effects greater at night due to the absence of photo-detachment process. The responses in the signal level due to Orionid meteor showers may also be governed by the time variation of the critical radius (r_c) under the combined influence of ambipolar diffusion, eddy diffusion and attachment but the contribution of r_c for Orionids should be examined elaborately for its model development and also to compare the time variation of r_c for other types of meteors like Leonids, etc. In addition, the spectrum monitoring stations could also be used for detecting radio transients and glowing radio sources in astronomy.

8.3 *Papers on which this book is based on*

1. Implementation of Radio Technique in Interstellar Radiation Field for Locating Distant Stars by A. B. Bhattacharya, D. K. Tripathi, A. Sarkar and A. Bhoumick, *International Journal of Applied Engineering Research,* Vol. 2, No 1, pp. 63-69, (2011).

2. A New Technique for Investigating Extrasolar Planets by A. B. Bhattacharya, S. Das, A. Sarkar, M. Debnath and A. Bhoumick, *International Journal of Physics,* Vol. 4, No 1, pp. 31-42, 2011.

3. Some Computations on the Drake Equation to Encapsulate the Probable Number of Broadcasting Civilizations by A. B. Bhattacharya, A. Sarkar and J. Pandit, *International Journal of Applied Engineering Research,* Vol. 2, No 2, pp. 590-954, 2011.

4. Implementation of Fast Fourier Transform for Detecting Solar Bursts by A. B. Bhattacharya, R. Bhattacharya, S. Das, D. Halder, A. Sarkar, A. Bhoumick, T. Das and M. Debnath, *Research Journal of Earth and Planetary Studies* Vol.1, No 1, pp. 009-018, 2011.

5. Electrification in the Earth and Martian Atmospheres with the Possibility of Global Electric Circuits by A. B. Bhattacharya, D. K. Tripathi, T. Das, A. Nag and A. Sarkar, *International Journal of Engineering Science and Technology (IJEST)*, Vol. 3, No 4, pp. 3221-3228, 2011.

6. VHF–Electromagnetic Signal Perturbations Related to Sikkim Earthquake of 18 September, 2011 by A. B. Bhattacharya, A. Sarkar, A. Bhoumick, J. Pandit, B. Raha and D. Halder, *International Journal of Theoretical And Applied Physics (IJTAP)*, Vol.2, No. II, pp. 65-72, December 2012.

7. Detection of Jovian Radio Bursts at High Altitudes by A.B. Bhattacharya, S. Mondal, J. Pandit, D. Halder, A. Sarkar and B. Raha, *International Journal of Engineering Science and Technology (IJEST)*, Vol. 4 No.06, 3029-3038, June 2012.

8. Orionid Meteor Shower and Its Effect on Ionospheric Propagation by A. B. Bhattacharya, J. Pandit, A. Bhoumick, D. Halder, A. Sarkar, B. Raha and R. Bhattacharya, *International Journal of Theoretical and Applied Physics (IJTAP)*, Vol.2, No. I, pp. 219-229, (June 2012).

9. Perturbations at 27 kHz Atmospherics in Relation to Indonesia Earthquake by A. B. Bhattacharya, B. Raha, A. Sarkar and J. Pandit, *International Journal of Applied Sciences and Engineering Research*, Vol. 1, No. 2, pp. 358- 364, 2012.

10. Estimation of the Frequency of Artifacts Likely to Fall Down to Earth and Other Plancts by A. B. Bhattacharya, J. Pandit, B. Raha, A. Sarkar, *International Journal of Modern Engineering Research (IJMER)*, Vol. 2, Issue. 3, pp. 722-727, 2012.

11. Transit of Venus Across the Sun at Kalyani Sky and Some Notable Changes in the Atmospherics Record by A. B. Bhattacharya, J. Pandit, B. Raha and A. Sarkar, *International J. of Engg. Research & Indu. Appls. (IJERIA),* Vol. 5, No. III, pp. 225-234, 2012.

12. Observation of Galactic Radio Signal Using Interstellar Hydrogen Line at 1420 MHz by A. B. Bhattacharya, **A. Sarkar**, B. Raha, J. Pandit and D. Halder, *International J. of Engg. Research & Indu. Appls. (IJERIA),* Vol. 5, No. IV, pp. 1-10, 2012.

13. Studying Transient Features through the Solar Corona by 406.7 MHz Radio Telescope by A. B. Bhattacharya, D. Halder, A. Sarkar, J. Pandit and B. Raha, *International J. of Multidispl. Research & Advcs. in Engg. (IJMRAE),* Vol. 4, No. IV pp. 225-234, October 2012.

14. Response of Earthquakes on Radio Signals having Distant Epicentres at Multi-Frequency Bands by A. B. Bhattacharya, A. Sarkar and J. Pandit, *International Journal of Electronics & Communication Technology (IJECT)*, Vol. 4, Issue Spl - 1, pp. 107-108, 2013.

15. Communication with Extraterrestrial Intelligence Through Radio Signals and Messages by A. B. Bhattacharya and A. Sarkar, *International Journal of Electronics & Communication Technology (IJECT)*, Vol. 5, Spl - 2, pp. 63-66, 2014.

FSC
www.fsc.org
MIX
Papier | Fördert gute Waldnutzung
FSC® C083411

Kolibri 360 GmbH
Ferdinand-Jühlke-Straße 7
99095 Erfurt, Deutschland
produktsicherheit@kolibri360.de

Druck:
CPI Druckdienstleistungen GmbH
im Auftrag der
Kolibri 360 GmbH
Ferdinand-Jühlke-Str. 7
99095 Erfurt